The Reaction Wheel Pendulum

The Reaction Wheel Pendulum
Daniel J. Block, Karl J. Åström, and Mark W. Spong

ISBN: 978-3-031-00699-9   paperback
ISBN: 978-3-031-01827-5   ebook

DOI 10.1007/978-3-031-01827-5

A Publication in the Springer series
SYNTHESIS LECTURES ON CONTROL AND MECHATRONICS #1

Lecture #1

Series Editor: Mark W. Spong, Coordinated Science Lab, University of Illinois at Urbana-Champaign

First Edition
10 9 8 7 6 5 4 3 2 1

The Reaction Wheel Pendulum

Daniel J. Block
College of Engineering Control Systems Lab
University of Illinois

Karl J. Åström
Department of Automatic Control, LTH
Lund Institute of Technology

Mark W. Spong
Coordinated Science Lab
University of Illinois

SYNTHESIS LECTURES ON CONTROL AND MECHATRONICS #1

ABSTRACT

This monograph describes the Reaction Wheel Pendulum, the newest inverted-pendulum-like device for control education and research. We discuss the history and background of the reaction wheel pendulum and other similar experimental devices. We develop mathematical models of the reaction wheel pendulum in depth, including linear and nonlinear models, and models of the sensors and actuators that are used for feedback control. We treat various aspects of the control problem, from linear control of the motor, to stabilization of the pendulum about an equilibrium configuration using linear control, to the nonlinear control problem of swingup control. We also discuss hybrid and switching control, which is useful for switching between the swingup and balance controllers. We also discuss important practical issues such as friction modeling and friction compensation, quantization of sensor signals, and saturation. This monograph can be used as a supplement for courses in feedback control at the undergraduate level, courses in mechatronics, or courses in linear and nonlinear state space control at the graduate level. It can also be used as a laboratory manual and as a reference for research in nonlinear control.

KEYWORDS

feedback control, inverted pendulum, modeling, dynamics, nonlinear control, stabilization, friction compensation, quantization, hybrid control.

Contents

CHAPTER 1

Introduction

"I THINK," shrilled Erjas, "that this is our most intriguing discovery on any of the worlds we have yet visited!"
— PENDULUM by Ray Bradbury and Henry Hasse, Super Science Stories, 1941

1.1   THE REACTION WHEEL PENDULUM

This monograph is concerned with modeling and control of a novel inverted pendulum device, called the *Reaction Wheel Pendulum*, shown in Figure 1.1. This device, first introduced in [18], is perhaps the simplest of the various pendulum systems in terms of its dynamic properties, consequently, its controllability properties. At the same time, the Reaction Wheel Pendulum exhibits several properties, such as underactuation and nonlinearity,[1] that make it an attractive and useful system for research and advanced education. As such, the Reaction Wheel Pendulum is ideally suited for educating university students at virtually every level, from entering freshman to advanced graduate students.

From a mechanical standpoint, the Reaction Wheel Pendulum is a *simple pendulum* with a rotating wheel, or bob, at the end. The wheel is attached to the shaft of a 24-Volt, permanent magnet DC-motor and the coupling torque between the wheel and pendulum can be used to control the motion of the system. The Reaction Wheel Pendulum may be thought of as a simple pendulum in parallel with a torque-controlled inertia (and therefore a double integrator).

This monograph can be used as a supplemental text and laboratory manual for either introductory or advanced courses in feedback control. The level of background knowledge assumed is that of a first course in control, together with some rudimentary knowledge of dynamics of physical systems. Familiarity with Matlab is also useful, as Matlab is used throughout as a programming environment. In subsequent chapters, we describe the dynamic modeling, identification, and control of the Reaction Wheel Pendulum and include suggested laboratory exercises illustrating important concepts and problems in control. Some

---

[1]We will define these terms and discuss them in detail in subsequent chapters.

FIGURE 1.1: The reaction wheel pendulum.

examples of the problems that can be easily illustrated using the Reaction Wheel Pendulum are

- Modeling
- Identification
- Simple motor control experiments: velocity and position control
- Nonlinear control of the pendulum
- Stabilization of the inverted pendulum
- Friction compensation
- Limit Cycle Analysis
- Hybrid control—swingup and balance of the pendulum.

1.2   THE PENDULUM PARADIGM

Taking its name from the Latin *pendere*, meaning to hang,[2] the pendulum is one of the most important examples in dynamics and control and has been studied extensively since the time of Galileo. In fact, Galileo's empirical study of the motion of the pendulum raised important questions in mechanics that were answered only with Newton's formulation of the laws of motion and later work of others. Galileo's careful experiments noted that a pendulum nearly returns to its released height and eventually comes to rest with lighter ones coming to rest

---

[2]Other cognates include suspend (literally, to hang below) and depend (literally, to hang from).

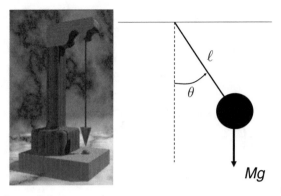

FIGURE 1.2: The simple pendulum.

faster. He discovered that the period of oscillation of a pendulum is independent of the bob weight, depending only on the pendulum length, and that the period is nearly independent of amplitude (for small amplitudes). All of these properties are now easily derived from Newton's laws and the equations of motion, discussed below.

In physics courses, therefore, the simple pendulum is often introduced to illustrate basic concepts like periodic motion and conservation of energy, while more advanced concepts like chaotic motion are illustrated with the forced pendulum and/or double pendulum.

1.2.1   The Simple Pendulum

To begin, let us consider a simple pendulum as shown in Figure 1.2 and discuss some of its elementary properties. Here, θ represents the angle that the pendulum makes with the vertical, ℓ and M are the length and mass, respectively, of the pendulum and g is the acceleration of gravity (9.8 m/sec^2 at the surface of the earth). The equation of motion of the simple pendulum is[3]

$$\ddot{\theta} + \frac{g}{\ell} \sin(\theta) = 0 \qquad (1.1)$$

Notice that the ordinary differential equation (1.1) is nonlinear due to the term $\sin(\theta)$. If we approximate $\sin(\theta)$ by θ, which is valid for small values of θ, we obtain the linear system

$$\ddot{\theta} + \omega^2 \theta = 0 \qquad (1.2)$$

where we have defined $\omega^2 := g/\ell$. Equation (1.2) is called the *simple harmonic oscillator*. One can verify by direct substitution that the above equation has the general solution

$$\theta(t) = A \cos(\omega t) + B \sin(\omega t) \qquad (1.3)$$

---

[3]Consult any introductory physics text.

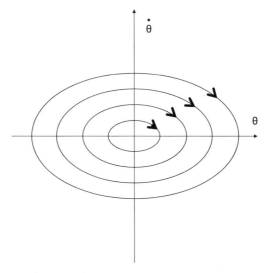

FIGURE 1.3: Phase portrait of the simple harmonic oscillator.

where A and B are constants determined by the initial conditions. In fact, it is an easy exercise to show that $A = \theta(0)$ and $B = \dot{\theta}(0)/\omega$. Moreover, the rate of change (velocity) of the harmonic oscillator is given by

$$\dot{\theta}(t) = B\omega \cos(\omega t) - A\omega \sin(\omega t) \tag{1.4}$$

and a directly calculation shows that

$$\theta^2(t) + \frac{1}{\omega^2}\dot{\theta}^2(t) = r^2 \tag{1.5}$$

where $r^2 = A^2 + B^2$. Equation (1.5) is the equation of an ellipse parameterized by time. This parameterized curve is called a *trajectory* of the harmonic oscillator system (see Figure 1.3). The totality of all such trajectories, one for each pair of initial conditions (A, B), is called the *phase portrait* of the system. Note that the period of oscillation of the simple harmonic oscillator (ω) is independent of the amplitude. It is surprising that, unlike the equation for the simple harmonic oscillator, which we easily solved, there is no closed form solution of the simple pendulum equation (1.1) analogous to Equation (1.3). A solution can be expressed in terms of so-called *elliptic integrals* but that subject is beyond the scope of this text. The phase portrait of the simple pendulum can be generated by numerical simulation as shown in Figure 1.4. We can gain added insight into this phase portrait by considering the scalar function

$$E = \frac{1}{2}\dot{\theta}^2 + \frac{g}{\ell}(1 - \cos(\theta)) \tag{1.6}$$

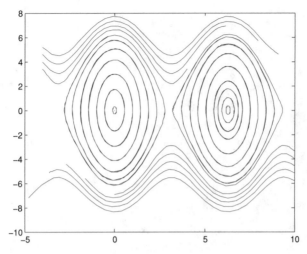

FIGURE 1.4: Phase portrait of the simple pendulum.

The function E is, in fact, proportional to the total energy, kinetic plus potential, of the simple pendulum. Computing the derivative of E yields

$$\dot{E} = \dot{\theta}\ddot{\theta} + \frac{g}{\ell}\sin(\theta)\dot{\theta} \qquad (1.7)$$

$$= \dot{\theta}\left(\ddot{\theta} + \frac{g}{\ell}\sin(\theta)\right) \qquad (1.8)$$

It follows that $\dot{E} = 0$ along solutions of the simple pendulum equation. This means that the function E is constant along solution trajectories, i.e., the level curves of E are trajectories. In classical mechanics, such a function is called a *first integral of the motion*. Each trajectory in Figure 1.4 is, therefore, a level curve of the function E. This is intuitively clear as we know from elementary physics that, without friction, the energy of the pendulum is constant. We will have much more to say about the pendulum dynamics in the chapters that follow. Since angles are typically given modulo 2π we obtain a nice representation by introducing cuts at $\pm\pi$ and glueing the parts together to form a cylinder. Such a representation, which is called a *manifold*, is useful when we do not have to take the number of rotations into account explicitly. Notice that for the particular implementation where the encoder is connected to wires it may be of interest to keep track of the number of revolutions so that we are not tearing the wires.

1.2.2 The Pendulum in Systems and Control

The simple pendulum system is interesting in its own right to study problems in dynamics and control. However, its importance is more than academic as many practical engineering

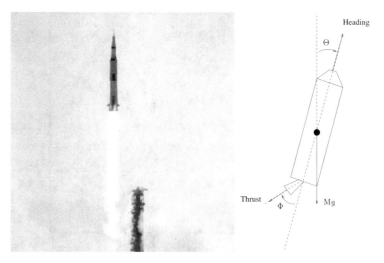

FIGURE 1.5: Liftoff of the Apollo 11 lunar mission. The rocket acts like an inverted pendulum balanced at the end of the thrust vectored motor.

systems can be approximately modeled as pendulum systems. In this section, we discuss several interesting examples and applications that can be modeled as pendulum systems.

Figure 1.5 shows the liftoff of a Saturn V rocket.[4] Active control is required to maintain proper attitude of the Saturn V rocket during ascent. Figure 1.5 also shows a diagram of a rocket whose pitch angle Θ can be controlled during ascent by varying the angle, Ψ, of the thrust vector. The pitch dynamics of the rocket can be approximated by a controlled simple pendulum.

In Biomechanics, the pendulum is often used to model bipedal walking. Figure 1.6 shows the Honda Asimo humanoid robot. In bipedal robots the stance leg in contact with the ground is often modeled as an inverted pendulum while the *Swing Leg* behaves as a freely swinging pendulum, suspended from the hip. In fact, studies of human postural dynamics and locomotion have found many similarities with pendulum dynamics. Measurements of muscle activity during walking indicate that the leg muscles are active primarily during the beginning of the swing phase after which they shut off and allow the leg to swing through like a pendulum. Nature has thus taught us to exploit the natural tendency of the leg to swing like a pendulum during walking, which may partially accounts for the energy efficiency of walking.

Likewise, quiet standing requires control of balance. So-called *Postural Sway* results from stretch reflexes in the muscles, which are a type of local feedback stabilization of the inverted pendulum dynamics involved in standing.

---

[4]This particular photo is of the Apollo 11 launch carrying astronauts Neil Armstrong, Buzz Aldrin, and Michael Collins on their historic journey to the moon.

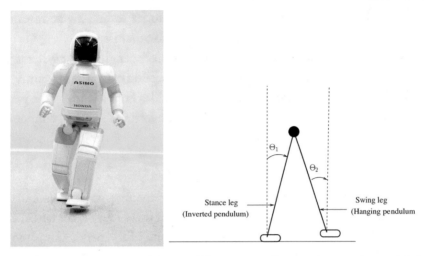

FIGURE 1.6: The Honda Humanoid Asimo. The swing and stance legs can be modeled as coupled pendula.

The *Segway Human Transporter*, shown on the left in Figure 1.7, is a recent invention that has achieved commercial success. The Segway is, in fact, a controlled inverted pendulum. Based on sensory input from gyros mounted in the base of the Segway, a computer control system maintains balance as the human rides it. The right side of Figure 1.7 shows an autonomous segway, designed to maintain balance and to locomote without human intervention. Such

FIGURE 1.7: The segway human transporter (left) and an autonomous self-balancing segway (right). Photos courtesy of the University of Illinois at Urbana–Champaign.

systems have been built at a number of research laboratories to investigate research issues in control, autonomous navigation, group coordination, and other issues.

There are many other examples of control problems in engineering systems where pendulum dynamics provides useful insight, including stabilization of overhead (gantry) cranes, roll stabilization of ships and trucks, and slosh control of liquids. Thus a study of the pendulum and pendulum-like systems is an excellent starting point to understand issues in nonlinear dynamics and control.

1.3 PENDULUM EXPERIMENTAL DEVICES

There have been several devices used to illustrate the dynamics of pendula and to facilitate control system design and implementation. The oldest of these ideas is the so-called *Cart-Pole* system shown in Figure 1.8. In this system, the pivot point of the pendulum is moved linearly in order to control the pendulum motion.

Later innovations of this idea were the *Pendubot* and the *Rotary* or *Furuta Pendulum*, shown in Figure 1.9. In both of these latter devices the second (or distal) link is a simple pendulum whose motion is controlled by the rotational motion (rather than linear motion) of

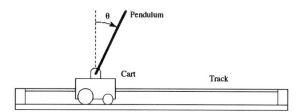

FIGURE 1.8: Cart-Pole system.

FIGURE 1.9: The pendubot (left) and the furuta pendulum (right).

the first (or proximal) link. The Pendubot is designed so that the axes of rotation of the two links are parallel while in the Rotary Pendulum, the axes of rotation are perpendicular.

The Reaction Wheel Pendulum is the newest and the simplest of the various pendulum experiments due to the symmetry of the wheel attached to the end of the pendulum. As we shall see in the next chapter on Modeling, this symmetry results in fewer coupling nonlinearities in the dynamic equations of motion and hence a simpler system to analyze, simulate, and control.

CHAPTER 2

Modeling

The first step in any control system design problem is to develop a mathematical model of the system to be controlled. In this section, we develop mathematical models for the Reaction Wheel Pendulum from first principles and we then perform some experiments to validate these models and to determine the parameters. A nonlinear model will first be derived using the Lagrangian approach. This model will later be linearized and the linear model used to design control strategies. In later sections, we will return to the nonlinear model and investigate the application of more advanced nonlinear control strategies for the problem of swingup control.

2.1 ANGLE CONVENTION AND SENSORS

A schematic diagram of the Reaction Wheel Pendulum is shown in Figure 2.1. The angle θ is the angle of the pendulum measured counterclockwise from the vertical when facing the system and θ_r is the wheel angle measured likewise. We have chosen the angles as in Figure 2.1 because it is natural to use gravity to line up the pendulum hanging down.

The Reaction Wheel Pendulum is provided with two optical encoders. These encoders are *Relative* as opposed to *Absolute* encoders and thus measure only the relative angle between

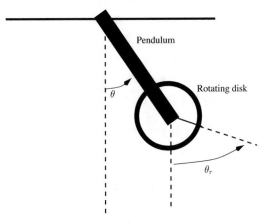

FIGURE 2.1: Schematic diagram of the Reaction Wheel Pendulum.

their (fixed) stator and (movable) rotor. Their values are initialized to zero at the start of any experiment. One encoder is attached to the fixed mounting bracket with its rotor shaft attached to the pendulum link. It thus provides a measure of the relative angle between the pendulum and the fixed base. The other encoder is attached to the motor fixed at the end of the pendulum. Its rotor shaft is attached to the rotating wheel and thus provides the relative angle between the pendulum and wheel. If we denote the encoder angles as φ and φ_r, respectively, then we see that

$$\theta = \varphi \tag{2.1}$$

$$\theta_r = \varphi + \varphi_r \tag{2.2}$$

Later we will discuss various issues, such as noise and quantization associated with the digital measurement of these angles and also the problem of estimating the angular velocities from the encoder values.

2.2   EQUATIONS OF MOTION

A convenient way to derive the equations of motion for mechatronic systems like the Reaction Wheel Pendulum is the so-called *Lagrangian method*. The Lagrangian method allows one to deal with scalar energy functions rather than vector forces and accelerations as in the Newtonian method and is, in many case, simpler.

The Reaction Wheel Pendulum has two degrees of freedom and we take as generalized coordinates the angles θ of the pendulum and θ_r of the rotor as shown in Figure 2.1. We introduce the following variables

m_p = mass of the pendulum
m_r = mass of the rotor
$m = m_p + m_r$ = combined mass of rotor and pendulum
J_p = moment of inertia of the pendulum about its center of mass
J_r = moment of inertia of the rotor about its center of mass
ℓ_p = distance from pivot to the center of mass of the pendulum
ℓ_r = distance from pivot to the center of mass of the rotor
ℓ = distance from pivot to the center of mass of pendulum and rotor

and, for later convenience, we define the following quantities

$$m = m_p + m_r$$
$$m\ell = m_p\ell_p + m_r\ell_r \tag{2.3}$$
$$J = J_p + m_p\ell_p^2 + m_r\ell_r^2$$

Lagrange's Equations

The Lagrangian method begins by defining a set of *generalized coordinates*, $q_1, \ldots, q_n$, to represent an *n*-degree-of-freedom system. These generalized co-ordinates are typically position coordinates (distances or angles).

In terms of these generalized coordinates, one then must compute the *kinetic energy*, T, and the *potential energy*, V. In general, the kinetic energy is a positive definite function of the generalized coordinates and their derivatives, while the potential energy is typically a function of only the generalized coordinates (and not their derivatives).

In a multi-body system, the kinetic and potential energies can be computed for each body independently and then added together to form the energies of the complete system. This is an important advantage of the Lagrangian method and works because energy is a scalar valued, as opposed to vector valued, function.

Once the kinetic and potential energies are determined, the *Lagrangian*, $L(q_1, \ldots, q_n, \dot{q}_1, \ldots, \dot{q}_n)$, is then defined as the difference between the kinetic and potential energies. The Lagrangian is, therefore, a function of the generalized coordinates and their derivatives.

The equations of motion are then expressed in terms of the Lagrangian in the following form,

$$\frac{d}{dt}\left(\frac{\partial L}{\partial \dot{q}_k}\right) - \frac{\partial L}{\partial q_k} = \tau_k, \quad k = 1, \ldots, n$$

The variable τ_k represents the generalized force (force or torque) in the q_k direction. These equations are called *Lagrange's equations*. For the class of systems considered here, Lagrange's equations are equivalent to the equations derived via Newton's second law.

The kinetic energy, T, of the system is the sum of the pendulum kinetic energy and the rotor kinetic energy and can be written in terms of the above quantities as

$$T = \frac{1}{2}J\dot{\theta}^2 + \frac{1}{2}J_r\dot{\theta}_r^2 \tag{2.4}$$

We assume that the potential energy, V, of the system is due only to gravity. Elasticity of the motor shaft or pendulum link would result in additional potential energy terms but we will assume that these effects are negligible. Thus the potential energy is

$$V = mg\ell(1 - \cos\theta) \tag{2.5}$$

where we have chosen to define the potential energy as being zero when the pendulum is hanging in the downward equilibrium. It is interesting to note that the potential energy does not depend on the rotor position since the mass of the rotor is distributed symmetrically about its axis of rotation.

The Lagrangian function, L, is then given by

$$L = T - V = \frac{1}{2}J\dot{\theta}^2 + \frac{1}{2}J_r\dot{\theta}_r^2 + mg\ell(\cos\theta - 1) \tag{2.6}$$

Taking the required partial derivative of the Lagrangian we find

$$\frac{\partial L}{\partial \dot{\theta}} = J\dot{\theta}, \qquad \frac{\partial L}{\partial \theta} = -mg\ell\sin\theta$$

$$\frac{\partial L}{\partial \dot{\theta}_r} = J_r\dot{\theta}_r, \qquad \frac{\partial L}{\partial \theta_r} = 0$$

In our case the torque produced by the motor results in a torque τ acting on the rotor and $-\tau$ acting on the pendulum. These are the two generalized forces in the θ_r and θ directions, respectively. Neglecting friction forces and the electrical dynamics of the DC-motor, the torque is given by

$$\tau = kI \tag{2.7}$$

where k is the torque constant of the motor and I is the motor current. Lagrange's equations are therefore

$$J\ddot{\theta} + mg\ell\sin\theta = -kI$$
$$J_r\ddot{\theta}_r = kI \tag{2.8}$$

The system given by Eq. (2.8) is characterized by four parameters: J, J_r, $mg\ell$, and k. However, dividing through by the moments of inertia, J and J_r, respectively, gives

$$\ddot{\theta} + \frac{mg\ell}{J}\sin\theta = -\frac{k}{J}I$$
$$\ddot{\theta}_r = \frac{k}{J_r}I \tag{2.9}$$

Thus the equations of motion are actually characterized by three parameters $mg\ell/J =: \omega_p^2$, k/J, and k/J_r. Notice that the parameter ω_p is the frequency of small oscillations of the system around the hanging position.

2.3 MODEL VALIDATION

Physical system modeling always involves trade-offs between accuracy and simplicity. That is, we would like the simplest model that still captures all of the important dynamic effects in the system. To derive the above model of the Reaction Wheel Pendulum we made several simplifying assumptions, for example, that elasticity in the pendulum link and motor shaft was negligible and that friction could be ignored.

Our first experiments are designed to investigate the validity of these modeling assumptions and to determine the parameters appearing in the equations of motion. We will first investigate the system when there is no control torque. It then follows from Eq. (2.9) that

$$\ddot{\theta} + \frac{mg\ell}{J}\sin\theta = 0$$

$$\ddot{\theta}_r = 0$$

(2.10)

Notice that the first equation is the equation for a pendulum with mass m, moment of inertia J, and center of mass at a distance ℓ from the pivot. Thus, if the pendulum is initialized at an angle θ_0 it will oscillate with constant amplitude. For small amplitudes the frequency of oscillation is $\omega_p = \sqrt{mg\ell/J}$. The second equation is simply a double integrator. If the angle θ_r and its derivative are zero the angle will remain zero for all times. One way to explore these equations is to investigate the motion when the pendulum is initialized at a given angle θ_0 with zero velocity and to investigate if θ will be periodic and θ_r will remain zero.

> **Experiment 1 (Simple Experiment with Free Swinging Pendulum).** Initialize the pendulum at an angle of about 20° let it swing and measure the pendulum and the rotor angles. Determine the frequency of the oscillation.

Figure 2.2 shows the measured pendulum angle θ and rotor angle θ_r for one such experiment. The behavior of the pendulum angle appears to be in reasonable agreement with the model. The rotor angle, however, is nearly identical to the pendulum angle, which is not predicted by the model. The reason for this is that there is friction between the pendulum and rotor. In effect, the rotor "sticks" to the pendulum and oscillates along with the pendulum.

To get more insight into what happens, we introduce the friction torques explicitly in the equations of motion. Equation (2.8) then becomes

$$J\ddot{\theta} + mg\ell\sin\theta = -kI - T_p + T_r$$

$$J_r\ddot{\theta}_r = kI - T_r$$

(2.11)

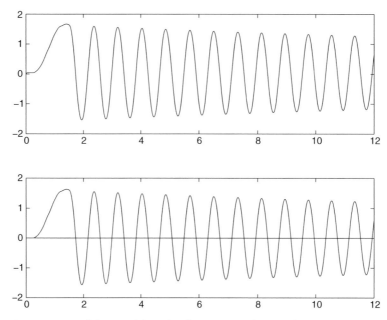

FIGURE 2.2: Free motion of the pendulum (top) and wheel (bottom) when no control is applied.

where T_p is the friction torque on the pendulum axis and T_r is the friction torque on the rotor axis. The torque T_p depends on the θ and $\dot{\theta}$ and the torque T_r depends on $\theta_r - \theta$ and $\dot{\theta}_r - \dot{\theta}$. The curves in Figure 2.2 indicate that the friction on the rotor axis is so large that the rotor is practically stuck to the pendulum. This implies that $\theta_r = \theta$. Adding the equations above we find

$$(J + J_r)\ddot{\theta} + mg\ell \sin\theta = -T_p$$

Notice that the friction torque on the rotor axis vanishes. This is very natural since there is no motion of the rotor relative to the pendulum. Also notice that, when the rotor and pendulum are stuck together and oscillate as a single mass, the frequency of small oscillations is

$$\omega_p' = \sqrt{\frac{mg\ell}{J + J_r}}$$

instead of ω_p.

More insight into the friction torque can be obtained by plotting the energy of the pendulum as a function of time. Expressions for the kinetic energy Eq. (2.4) and potential energy Eq. (2.5) were already obtained when deriving the equations of motion. If the rotor is

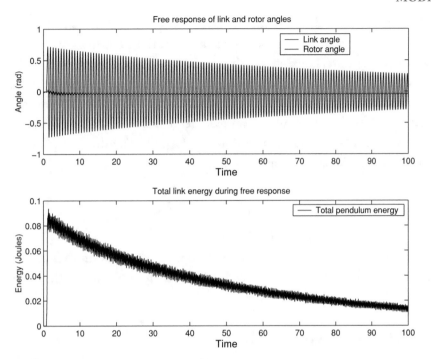

FIGURE 2.3: Total energy versus time for the data in Figure 2.2.

fixed to the pendulum the energy becomes

$$E = \frac{1}{2}(J + J_r)\dot{\theta}^2 + mg\ell(1 - \cos\theta)$$

Figure 2.3 shows the total energy as a function of time for the data in Figure 2.2. The energy decay due to friction T_p at the pendulum axis is quite small. Figure 2.3 indicates a time constant of nearly 1 min. For this reason we will ignore the pendulum friction in the subsequent modeling but we will model the rotor friction since it has a significant influence on the system. Before doing this we will go ahead and make estimates of the parameters of the model.

2.4   THE MOTOR DYNAMICS
The dynamics of the permanent-magnet DC motor can be written as

$$L\frac{dI}{dt} + RI = V - k\omega \qquad (2.12)$$

where L and R are the armature inductance and resistance, respectively, k is the motor back EMF constant (which is identical to the torque constant in mks-units) and V is the applied voltage. The electrical time constant of the motor is $L/R = 0.0005$.

With a current-controlled DC motor the applied voltage is changed to compensate for the back EMF. It follows from the above equation that the compensation required increases with the velocity. There are, however, physical limits to what can be achieved. If the maximum voltage of the drive amplifier is V_{max} it follows from the above equation that the current feedback will cease to function if the rotor velocity is sufficiently large. Neglecting dynamics in Eq. (2.12) we find that to obtain a positive current we must require that $\omega \le V_{max}/k$ and similarly that a negative current can be generated only if $\omega \ge -V_{max}/k$. The drive amplifier with current feedback will thus only function as intended if

$$-\frac{V_{max}}{k} \le \omega \le \frac{V_{max}}{k} \tag{2.13}$$

No torque is generated if this inequality is not satisfied.

The current I, which we have taken as the control input, is thus filtered by the motor with a time constant of 0.0005 s. There may be additional dynamics due to the current feedback loop.

2.5   THE DRIVE AMPLIFIER

The motor current is generated by a pulse width modulation system. The basic cycle is 20 kHz. Each cycle is divided into 500 segments and the control signal sets the duty cycle. The pulse width modulator is controlled from the computer. Because of the current feedback, the current is proportional to the control command, u, from the computer. The control variable used in the computer is scaled so that 10 units correspond to maximum current. Therefore we can write

$$kI = k_u u ; \quad |u| \le 10 \tag{2.14}$$

where the proportionality constant k_u satisfies

$$k_u = \frac{k I_{max}}{10} = 0.00493$$

An independent calibration of the torque constant k_u can be made using a mechanical torque meter and plotting the torque as a function of the current.

Experiment 2 (Determination of Static Torque Characteristics).   For this experiment you need a torque meter. Connect this to the rotor axis. Use the computer to apply a current to the rotor amplifier. Measure the torque for different values of the control signal and plot the results.

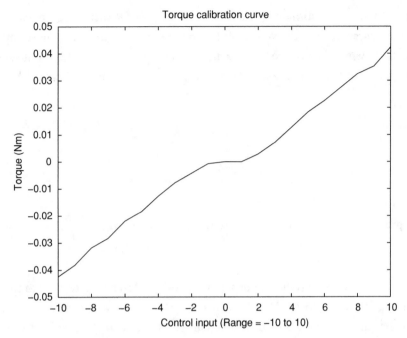

FIGURE 2.4: Measured values of torque as a function of the control signal u. Maximum torque corresponds to $u = 10$.

The results of such an experiment are shown in Figure 2.4. Notice that the curve has a dead-zone at the origin because of friction. Fitting straight lines to the linear portions of the curve we get $k_u = 0.00494$ which agrees well with the value computed above.

2.6   DETERMINATION OF PARAMETERS

Having found that the model is reasonable even if it is not perfect we will now determine the parameters. The parameters can be determined from physical construction data and by direct experiments on the system. It is useful to combine both methods to make cross checks.

Since the friction torque on the pendulum axis is small we will neglect it. The brushes in the motor are the main contributors to the friction torque on the motor axis. To start with it will be assumed that it can be modeled as Coulomb friction, i.e., a constant torque that is in the opposite direction of the motion. From the curve in Figure 2.4 we find that approximately one unit of control is required to make the system to move. This means that the Coulomb friction torque is approximately 0.005 Nm. It follows from the data sheet for the motor that this is twice the friction torque of the motor without load.

Having obtained an estimate of the friction torque on the rotor axis we will now proceed to determine the parameters of the system. It is convenient to use the normalized representation given by Eq. (2.9) which is close to physics and has few parameters. The input will, however, be chosen as the variable u in the computer that represents the control signal. The model then becomes.

$$\frac{d^2\theta}{dt^2} + a \sin\theta = -b_p(u - F)$$

$$\frac{d^2\theta_r}{dt^2} = b_r(u - F)$$

(2.15)

where F represents the friction torque on the motor axis, $a := \dfrac{mgl}{J}$, $b_p := \dfrac{k_u}{J} = \dfrac{kI_{max}}{10J}$, $b_r := \dfrac{k_u}{J_r}$.

Note that the friction torque depends on the motion of the rotor relative to the pendulum. It is convenient, however, to express it as in Eq. (2.15) because the friction torque is then expressed in the same units as the control signal. If we want to convert it to Nm we simply multiply by the value by k_u. The value of F is approximately 1. Later we will show that F depends on the angular velocity.

By measuring the dimensions of the components, weighing them and computing moments of inertia using simplified formulas we find.

$$m_p = 0.2164\,\text{kg} \quad J_p = 2.233\,10^{-4}\,\text{kg}\,\text{m}^2 \quad \ell_p = 0.1173\,\text{m}$$

$$m_r = 0.0850\,\text{kg} \quad J_r = 2.495\,10^{-5}\,\text{kg}\,\text{m}^2 \quad \ell_r = 0.1270\,\text{m}$$

From these values we obtain

$$m = 0.3014\,\text{kg}$$

$$\ell = 0.1200\,\text{m}$$

$$J = 4.572\,10^{-3}\,\text{kg}\,\text{m}^2$$

$$\omega_p = \sqrt{\frac{mg\ell}{J}} = 8.856\,\text{rad/s}$$

$$\omega'_p = \sqrt{\frac{mg\ell}{J + J_r}} = 8.832\,\text{rad/s}$$

The data sheet for the motor, Pittman LO-COG 8 × 22, gives the following values

$k = 27.4 \times 10^{-3}$ Nm/A	Motor Torque Constant
$R = 12.1$	Armature Resistance
$I_{max} = 1.8$ A	Maximum Motor Current
$L = 0.00627$	Armature Inductance
$T_{peak} = 38.7 \times 10^{-3}$ Nm	Maximum Motor Torque
$T_e = 0.5$ ms	Electrical Time Constant
$\omega_{max} = 822$ rad/s	Maximum Motor Speed
$V_{max} = 22$ V	Maximum Motor Voltage

Performing the indicated calculations, we find that the parameters of the model are

$$a = \omega_p^2 = 78.4$$
$$b_p = 1.08$$
$$b_r = 198$$

The controller parameters can be computed with the Matlab program shown at the end of this chapter. We can obtain a cross check by determining the parameters experimentally. The parameter a was already obtained in the model validation. The parameters b_p and b_r can be obtained from the following experiment. Apply a control signal u_0 for a short time h. If h is sufficiently small the sinusoidal term in Eq. (2.15) can be neglected and the equation becomes

$$\ddot{\theta} = -b_p(u - F)$$
$$\ddot{\theta}_r = b_r(u - F)$$

Both angles will then change quadratically during the interval $0 \le t \le h$ with rates given by the parameters b_p and b_r. At time $t = h$ we have

$$\dot{\theta}(h) \approx -b_p(u_0 - F)h$$
$$\dot{\theta}_r(h) \approx b_r(u_0 - F)h$$

where F is the control signal required to compensate for friction. The velocities will then remain constant. It is useful to repeat the experiment for different values of the control signal to investigate if the system is linear. To make sure that the sinusoidal term is negligible the pulse width should be chosen so that $\omega_p h$ is small.

Experiment 3 (Applying a Torque Pulse to the System). Apply a torque pulse to the system as described above. Measure the pendulum and rotor angles as functions of time. Use the results to determine the coefficients b_p and b_r.

Summary

In this chapter, we have found that the Reaction Wheel Pendulum can be described by the model (2.15)

$$\ddot{\theta} + a \sin \theta = -b_p(u - F)$$
$$\ddot{\theta}_r = b_r(u - F)$$

The angles are given in radians, the control signal u is the control signal used in the computer and is constrained to lie in the range ± 10. One unit of u corresponds to a torque of 0.0005 Nm. The variable F represents the friction torque on the rotor axis. The friction F is in the range of 1 to 2 torque units.
The parameters have the values

$$a = 78$$
$$b_p = 1.08$$
$$b_r = 198$$

There are additional dynamics because of the motor time constant which is of the order of 0.5 ms. There is also an additional delay in the sensing, which depends on the sampling period used in the computer. This will be discussed more in the next chapter.
We note that the system is nonlinear, but approximately linear for small values of θ. In addition to the nonlinear gravitational force acting on the pendulum there are additional nonlinearities in the system caused by friction and saturation of the amplifiers. The effects of friction will be discussed later. The saturation effects are caused by the limited voltage of the drive amplifier and the back EMF. The net effect is that no torque will be generated by the rotor if the inequality (2.13) is violated.
It follows from Eq. (2.12) that the maximum motor velocity is given by

$$\omega_{max} = \frac{V_{max}}{k_m} = \frac{22.7}{0.00274} = 828 \text{ rad/s}.$$

An alternative would be to include the dynamics of the motor current given by Eq. (2.12) with the current feedback in the model and introduce a limit on the voltage. This would make the model more complicated. Since we are interested in controlling the pendulum which has a natural frequency of about 9 rad/s and the electrical time constant of the motor is of the order of 0.5 ms we have chosen to use the simple model and treat (2.12) as unmodeled dynamics.

```
%MATLAB Program that enters the system parameters
%and computes the model parameters    .
g=9.91
mp=0.2164
mr=0.0850
lp=0.1173
lr=0.1270´
Jp=2.233e-4
Jpe=mp*lp*lp/12
Jr=2.495e-5
%Derived data
J=Jp+mp*lp*lp+mr*lr*lr
m=mp+mr
l=(mp*lp+mr*lr)/m
wp=sqrt(m*g*l/J)
wp1=sqrt(m*g*l/(J+Jr))
%Pittman LO-COG 8X22
km=27.4e-3
Imax=1.8
R=12.1
L=6.27e-3
Vmax=22.7
Te=L/R
wmax=Vmax/km
ke=0.00494 %Nm per unit
kee=Imax*km/10
kt=0.00494
b1=km/J
b2=km/Jr
a=wp^2
bp=ke/J
br=ke/Jr
```

CHAPTER 3

Controlling the Reaction Wheel

We will begin our study of control by first controlling only the reaction wheel. To do this we will reconfigure the system by removing the pendulum and attaching the motor and wheel directly to the mounting bracket as shown in Figure 3.1.

The equation of motion of the rotor is

$$\frac{d^2\theta_r}{dt^2} = b_r(u - F) \qquad (3.1)$$

This model is just a double integrator and therefore easy to control. The model (3.1) is a standard model used in velocity and position control of many mechatronic systems.

3.1 POSITION SENSING

The control of the wheel is complicated by the fact that we have only a digital measurement of position available. In other words, the optical encoder on the motor provides a measurement of the wheel angle at a resolution of 4000 counts/revolution. This means that the smallest change

FIGURE 3.1: Configuration of the system used to make experiments with wheel control.

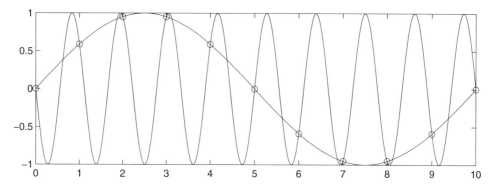

FIGURE 3.2: Uncertainty introduced by sampling.

in angle that can be sensed is

$$\Delta_\theta = \frac{2\pi}{4000} = 0.00157 \text{ radians}$$

or about $0.900°$. As the wheel rotates, there will be a ripple in the angle signal corresponding to this value. This effect is known as the *Quantization Error* and limits the achievable accuracy in control. Obviously, we cannot hope to reduce the error in the wheel angle below that which we can measure, i.e., below the sensor resolution.

An additional source of uncertainty in our knowledge of the wheel position results from *Sampling Error*. Since position measurements are recorded only at discrete instants of time, once during every sampling interval, we have no measurement of what happens between sample times. Figure 3.2 illustrates the effect of discrete sampling of a continuous quantity. In this figure, two different signals lead to exactly the same set of digital measurements. Although we will not discuss it further here, it is well known that one must sample at least as fast as the so-called *Nyquist Frequency* in order to reconstruct uniquely a continuous, band-limited signal. (The Nyquist frequency is twice the maximum frequency present in the signal.) In our experiments we will sample sufficiently fast, relative to the maximum speed of the motor, that this phenomenon, known as **Aliasing**, will not pose a significant problem for us.

3.2   VELOCITY ESTIMATION

Since there is no direct measurement of the wheel velocity, we will have to estimate or compute the velocity from the position measurements. This will introduce error and uncertainty into the velocity. In this section, we discuss different ways to estimate the velocity from discrete measurements of position.

The simplest way to estimate the velocity from position measurements is just to compute the *angle difference* over the sample interval by taking the difference between consecutive angle measurements.

$$\omega_k = \frac{\theta_k - \theta_{k-1}}{h} \qquad (3.2)$$

where θ_k represents the kth encoder sample (in radians) and h is the sampling interval.

The resolution in velocity is thus

$$\Delta_\omega = \frac{\Delta_\theta}{h} = \frac{0.00157}{h}$$

Note that the resolution increases with decreasing sampling period. With $h = 0.001$, the velocity resolution is 1.57 rad/s. When the velocity changes there will also be a ripple in the velocity signal corresponding to Δ_ω. This ripple will be amplified by the control signal.

3.3   FILTERING THE VELOCITY SIGNALS

The ripple in the velocity signal due to sampling generally has a significant effect on the performance of the control system. It is therefore of interest to filter the velocity signal to try to remove this ripple. Assume that we would like a first-order filter with the input/output relation

$$Y_f(s) = \frac{1}{1 + sT} Y(s)$$

where Y and Y_f represent the measured and filtered signals, respectively. Hence

$$s\, TY_f(s) + Y_f(s) = Y(s)$$

This corresponds to the differential equation

$$T\frac{dy_f}{dt} + y_f = y$$

Approximating the derivative with a difference we get

$$T\frac{y_f(t) - y_f(t - h)}{h} + y_f(t) = y(t)$$

Hence

$$y_f(t) = \frac{T}{T + h} y_f(t - h) + \frac{h}{T + h} y(t)$$

With this filter the high-frequency error is reduced by the factor $h/(T + h)$. Notice that the filter will introduce additional dynamics in the loop. The ripple in the filtered velocity signal caused by the encoder is

$$\Delta_{y_f} = \frac{h}{T+h}\frac{\Delta_\theta}{h} = \frac{0.00157}{T+h} \tag{3.3}$$

Assume for example that the sampling period is 0.5 ms and the filtering time constant is $T = 2.5$ ms. Then the filtering equation becomes

$$y_f(t) = \frac{5}{6}y_f(t-h) + \frac{1}{6}y(t)$$

The ripple in the filtered velocity is

$$\Delta y_f = \frac{0.00157}{0.003} = 0.52$$

Experiment 4 (Velocity Filter). Set up the motor/wheel system so that you can apply an open loop control signal to spin the motor at various speeds. Record the signal from the encoder. From these encoder data, compute the average velocity by taking the difference in position divided by the sample time. Then implement the first-order filter described in this section and experiment with various values of the cutoff frequency. Plot the filtered velocity signals and compare with the velocity computed by the averaging method.

Figure 3.3 shows the results of one such experiment. Note that the velocity generated from the first-order filter is smoother because the filter has the effect of "cutting off" or attenuating the high-frequency component of the measured signal.

3.4 VELOCITY OBSERVER

The velocity estimation methods in the previous section utilized only the encoder and timing data but did not utilize the system model (3.1). We might conjecture that utilizing knowledge of the system dynamics could lead to a more accurate velocity estimate. We will explore this conjecture in this section.

Estimators that incorporate the plant dynamics to estimate the state variables are known as *Observers*. To begin we write the model equations (3.1) in state space form as

$$\begin{aligned}\dot{x}_1 &= x_2\\ \dot{x}_2 &= b_r(u - F)\end{aligned} \tag{3.4}$$

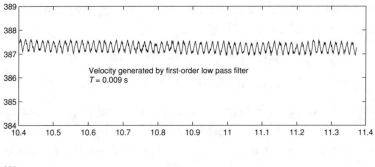

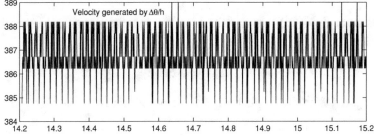

FIGURE 3.3: Comparison of average velocity versus velocity generated by a first-order filter.

where $x_1 = \theta_r$ and $x_2 = \dot{\theta}_r$. An observer for this system can be written as

$$\frac{d\hat{x}}{dt} = \begin{pmatrix} 0 & 1 \\ 0 & 0 \end{pmatrix} x + \begin{pmatrix} 0 \\ b_r \end{pmatrix} (u - F) + \begin{pmatrix} k_1 \\ k_2 \end{pmatrix} (y - \hat{x}_1) \qquad (3.5)$$

Introducing the error $e = x - \hat{x}$ and subtracting Eqs. (3.5) from (3.4) we find that the observer error is given by

$$\frac{de}{dt} = \begin{pmatrix} -k_1 & 1 \\ -k_2 & 0 \end{pmatrix} e$$

This equation has the characteristic polynomial

$$s^2 + k_1 s + k_2$$

The error will go to zero if the filter gains k_1 and k_2 are positive. Choosing the filter gains so that the characteristic polynomial is

$$s^2 + 2\zeta\omega_o + \omega_o^2$$

we find that the gains are given by

$$k_1 = 2\zeta\omega_o$$
$$k_2 = \omega_o^2$$

To implement the observer on the computer we simply approximate the derivatives with differences and we get the following difference equation

$$\hat{x}_1(t + h) = \hat{x}_1(t) + h\hat{x}_2(t) + hk_1(y(t) - \hat{x}_1(t))$$
$$\hat{x}_2(t + h) = \hat{x}_2(t) + hb_r(u(t) - F(t)) + hk_2(y(t) - \hat{x}_1(t)) \tag{3.6}$$

The observer requires the friction torque F. Since this is unknown, we will try to neglect it by setting $F = 0$ in the observer. Noticed that the filtered velocity is obtained by combining the measured angle and the current fed to the motor. The fact that the current is used provides phase lead.

It follows from Eq. (3.6) that the ripple in the velocity signal from the observer caused by the encoder is

$$\Delta_{\hat{x}_2} = k_2 h \Delta_\theta \tag{3.7}$$

This can be compared with the ripple for the velocity estimate obtained from the filtered angle differences

$$\Delta_{y_f} = \frac{\Delta_\theta}{T + h}$$

See Eq. (3.3). The ripple from the filters are the same if

$$k_2 = \frac{1}{h(T + h)}$$

with $h = 0.001$ and $T = 0.009$ we get $k_2 = 100000$. If k_2 is smaller than this value the observer gives a velocity estimate with less ripple than the filtered angle difference.

Experiment 5 (Comparison of a Simple Velocity Filter with an Observer). Using the data generated in Experiment 4, compute the estimated velocity using the discrete-time observer described above. Compare with the results obtained with the simple velocity filter.

Figure 3.4 shows the outputs of the filtered angle difference and the observer when the motor is running at constant speed. Notice that there is a difference between the outputs. The velocity generated by the second-order observer shows a steady state error, i.e., an offset from the filtered velocity. The reason for this difference is that the friction force was neglected in the design of the model-based observer. To understand this we will investigate the consequences of neglecting the friction force.

In the experiment the input signal is constant $u = u_0$. Assume that the model has a constant friction force F_0 but that the friction force is neglected in the observer. The error

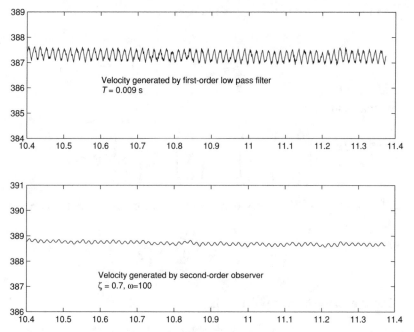

FIGURE 3.4: Observer estimate (bottom) and first-order filter (top).

equation then becomes

$$\frac{de}{dt} = \begin{pmatrix} -k_1 & 1 \\ -k_2 & 0 \end{pmatrix} e - \begin{pmatrix} 0 \\ b_r \end{pmatrix} F$$

The steady state error is given by

$$e = \begin{pmatrix} -k_1 & 1 \\ -k_2 & 0 \end{pmatrix}^{-1} \begin{pmatrix} 0 \\ b_r \end{pmatrix} F = \frac{1}{k_2} \begin{pmatrix} 0 & -1 \\ k_2 & -k_1 \end{pmatrix} \begin{pmatrix} 0 \\ b_r \end{pmatrix} F = - \begin{pmatrix} \dfrac{b_r}{k_2} \\ \dfrac{b_r k_1}{k_2} \end{pmatrix} F$$

The friction force will thus give steady state errors in the estimates. We will see later that the friction force F at a constant motor speed of 225 rad/s is approximately 3.7 Nm. Evaluating the numerical value of the velocity estimate in the experiment we find that

$$e_2 = x_2 - \hat{x}_2 = -\frac{b_r k_1}{k_2} F = 10.36 \text{ rad/s}$$

which agrees well with the experiments.

To obtain a good velocity estimate it is thus necessary to consider the friction torques. A simple approach is to assume that the friction is constant and introduce $b_r F$ as an extra state variable x_3. The model (3.4) for the system then becomes.

$$\dot{x}_1 = x_2$$
$$\dot{x}_2 = x_3 + b_r u \qquad (3.8)$$
$$\dot{x}_3 = 0$$

An observer for this system is

$$\frac{d\hat{x}}{dt} = \begin{pmatrix} 0 & 1 & 0 \\ 0 & 0 & 1 \\ 0 & 0 & 0 \end{pmatrix} x + \begin{pmatrix} 0 \\ b_r \\ 0 \end{pmatrix} u + \begin{pmatrix} k_1 \\ k_2 \\ k_3 \end{pmatrix} (y - \hat{x}_1) \qquad (3.9)$$

Subtracting (3.9) from (3.8) gives the following equation for the estimation error $e = x - \hat{x}$.

$$\frac{de}{dt} = \begin{pmatrix} -k_1 & 1 & 0 \\ -k_2 & 0 & 1 \\ -k_3 & 0 & 0 \end{pmatrix} e$$

This equation has the characteristic polynomial

$$s^3 + k_1 s^2 + k_2 s + k_3$$

Equating the coefficients of equal powers of s with the standard third-order polynomial

$$(s + \alpha \omega_0)(s^2 + 2\zeta \omega_0 s + \omega_0^2)$$

we find that the filter gains are given by

$$k_1 = (\alpha + 2\zeta)\omega_0 \quad k_2 = (1 + 2\alpha\zeta)\omega_0^2$$
$$k_3 = \alpha\omega_0^3 \qquad (3.10)$$

It is natural to associate the mode $\alpha\omega_0$ with estimation of the friction.

A discrete-time version of the observer is obtained by replacing the derivatives by differences. This gives

$$\hat{x}_1(t + h) = \hat{x}_1(t) + h\hat{x}_2(t) + hk_1(y(t) - \hat{x}_1(t))$$
$$\hat{x}_2(t + h) = \hat{x}_2(t) + h\hat{x}_3(t) + hb_r u(t) + hk_2(y(t) - \hat{x}_1(t)) \qquad (3.11)$$
$$\hat{x}_3(t + h) = \hat{x}_3(t) + hk_3(y(t) - \hat{x}_1(t))$$

Experiment 6 (Augmented Observer).   Using the data generated in Experiment 4, compute the estimated velocity using the third-order augmented observer described above. Compare with the results obtained with the simple velocity filter and the experiment with the second-order observer.

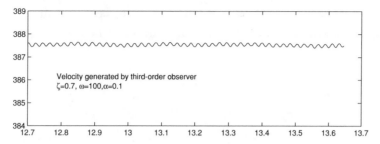

FIGURE 3.5: The third-order observer.

Figure 3.5 shows the results of one such experiment with the third-order observer. Note that the steady state error in wheel angular velocity has been removed and that the velocity signal is also much smoother than that generated by either the averaging filter or the first-order filter.

3.5   VELOCITY CONTROL

We now have signals available that provide estimates of both the wheel position and velocity. These signals can be used for feedback control. In this section, we will investigate control of the wheel speed and later control of the wheel position.

We will first make the wheel spin at constant rate. Let the angular velocity of the wheel be $\omega = d\theta_r/dt$. It follows from Eq. (3.1) that

$$\frac{d\omega}{dt} = b_r(u - F) \qquad (3.12)$$

The proportional feedback

$$u = k_{dr}(\omega_r - \omega) \qquad (3.13)$$

gives a closed loop system characterized by

$$\frac{de}{dt} + k_{dr}b_r e = b_r F$$

where $e = \omega_r - \omega$ is the error. With the chosen control law the angular velocity will follow the reference value ω_r, with a steady state error.

$$e_{ss} = \frac{F}{k_{dr}}$$

The time constant of the closed loop system is

$$T_\omega = \frac{1}{k_{dr}b_r}$$

The response time of the system will decrease with increasing feedback gain k_{dr}. It is interesting to see how large the feedback gain can be made or equivalently how fast the closed loop system can be made.

> **Experiment 7 (Velocity Control with Proportional Feedback).** Program the velocity control law $u = k_{dr}(\omega_r - \omega)$. Investigate the step response for various values of the gain k_{dr} and various values of the reference speed, ω_r. Notice which values of the gains and reference values agree with the predicted response. Try to explain any deviations from the predicted responses.

In the above experiment you should notice that there will be a limit to the speed of response achievable as you increase the gain. To investigate this further we have to make a more accurate model taking into account the electrical time constant of the motor, which was found to be around 0.5 ms, and the effect of the sampling delay. Estimating the velocity by taking differences of the encoder signal, for example, introduces a delay of half a sampling interval. Approximating the time delay with a time constant, and lumping all dynamics into one time constant, we find that the system has an additional time constant

$$T_e = 0.0005 + \frac{h}{2}$$

with a sampling period of 1 ms we find that the additional time constant is 1 ms.

Taking the additional dynamics into account we find that the closed loop system has the characteristic polynomial

$$s(s\,T_e + 1) + k_{dr}b_r = T_e\left(s^2 + \frac{s}{T_e} + \frac{k_{dr}b_r}{T_e}\right)$$

The relative damping is

$$\zeta = \frac{1}{2\sqrt{b_r k_{dr}\,T_e}}$$

Requiring that the relative damping is greater than 0.707 we get the inequality

$$k_{dr}b_u\,T_e < 0.5$$

Inserting the numerical values we get $k_{dr} < 5$ for infinitely fast sampling. With a sampling period of 1 ms we get $k_{dr} < 2.5$. Notice that the admissible gain decreases with increasing sampling period. We could try to increase the gain more by using a controller with derivative action.

3.6   PI CONTROL

We have found that a friction torque F gives a steady state error in the velocity of F/k_{dr}. To reduce the effects of friction it is, therefore, desirable to have a high gain of the velocity controller, but we have also found that the additional dynamics gives limitations to the controller gain. Integral action can be used to eliminate the steady state error without requiring large gains. Introducing an integral control term, the control law becomes

$$u = k_{dr}(\omega_r - \omega) + \frac{k_i}{h} \int_0^t (\omega_r - \omega(\tau))d\tau \qquad (3.14)$$

Inserting this control law into (2.15) gives a closed loop system with the characteristic polynomial

$$s^2 + b_r k_{dr} s + b_r k_i$$

Identifying coefficients of equal powers of s with the characteristic polynomial of the standard second-order system

$$s^2 + 2\zeta \omega_0 s + \omega_0^2$$

we find that the controller parameters can be expressed as

$$k_{dr} = \frac{2\zeta \omega_0}{b_r} \qquad (3.15)$$

$$k_i = \frac{\omega_0^2 h}{b_r} \qquad (3.16)$$

Experiment 8 (PI Speed Control). Design and test a PI controller for the above system. Parameterize the gains k_{dr} and k_i as in Eq. (3.16). Start with a fixed value of ζ, say $\zeta = 1$, which gives critical damping, and experiment with different values of ω_0 to give the desired response speed.

3.7   FRICTION MODELING

Friction is a complicated phenomena. The friction force depends on many factors, for example, the relative velocity at the friction surface. Having obtained controllers for the wheel velocity the dependence of friction on velocity can be determined. To do this we use the velocity controller to run the wheel at constant speed. The control signal required to do this is then equal to the friction torque.

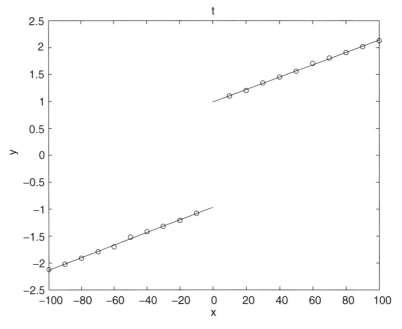

FIGURE 3.6: Control signal u as a function of the angular velocity of the wheel. With proper scaling this is the friction curve of the wheel.

Experiment 9 (Determination of Friction Curve).   Connect the system with PI control of the velocity. Adjust the controller parameters to give a good response. Determine the average control signal for a given velocity. Repeat the experiment for different velocities, both positive and negative, and plot the control signal as a function of the velocity.

Figure 3.6 shows the results of such an experiment. The figure shows that the friction torque has a Coulomb friction component and a component which is linear in the velocity. Fitting straight lines to the data in the figure gives the following model for the friction.

$$F = \begin{cases} 0.99 + 0.0116\omega, & \text{if } \omega > 0 \\ -0.97 + 0.0117\omega, & \text{if } \omega < 0 \end{cases} \qquad (3.17)$$

3.8   FRICTION COMPENSATION

Having obtained a reasonable friction model we can now attempt to compensate for the friction. This is done simply by measuring the velocity and adding a term given by the friction model (3.17) to the control signal.

Experiment 10 (Friction Compensation).   Use the results of Experiment 9 to design a control strategy that compensates for friction. Implement this control law in such a way that it can be switched on and off. Spin the wheel manually with and without the friction compensator and observe the differences.

3.9   CONTROL OF THE WHEEL ANGLE

We next consider control of the wheel angle. To do this we introduce the controller

$$u = k_{pr}(\theta_{ref} - \theta_r) - k_{dr}\dot{\theta}_r \qquad (3.18)$$

Substituting this controller into the linearized equation of motion of the system (3.1) we find that the closed loop system is described by

$$\ddot{\theta}_r + k_{dr}b_r\dot{\theta}_r + k_{pr}b_r\theta_r = k_{pr}b_r\theta_{ref} - b_r F \qquad (3.19)$$

Comparing this with the equation for a mass, spring, damper system

$$m\frac{d^2x}{dt^2} + d\frac{dx}{dt} + kx = kx_0$$

we find that the damping term and the stiffness term can be set directly by the feedback. Using feedback it is thus possible to obtain behavior equivalent to that obtained by adding springs and dampers. Feedback is more convenient because it gives great flexibility in modifying the apparent damping and stiffness parameter values.

The closed loop system has the characteristic polynomial

$$s^2 + k_{dr}b_r s + k_{pr}b_r$$

Identifying the coefficients of equal powers of s with the standard second-order polynomial

$$s^2 + 2\zeta\omega_0 s + \omega_0^2$$

we find

$$k_{pr} = \frac{\omega_0^2}{b_r}$$

$$k_{dr} = \frac{2\zeta\omega_0}{b_r}$$

Large values of ω_0 give high controller gains. Small disturbances and sensor noise will then be amplified and they will generate large control signals. The model we have used is only valid in a certain frequency range. For higher frequencies there are other phenomena that must be accounted for. It is interesting to investigate experimentally how fast the system can be made. This is easily done by observing the behavior of the system when ω_0 is increased.

Experiment 11 (Control of Wheel Angle). Implement the controller (3.18) in such a way that the controller is parameterized in ζ and ω_0. Determine experimentally how large the value of ω_0 can be made. Compare the results with the time constant of the filtering and other neglected time constants. Also, investigate the effect of sensor noise.

Summary

In this chapter, we have investigated the control of the wheel without consideration of the pendulum. We first considered the problem of sensing. We studied three ways to estimate the wheel velocity using the position data from the optical encoder attached to the motor

1. by computing the *Average Velocity* over a sample period
2. by a *First-Order Low Pass Filter*
3. by a model-based *Observer*

We found that the Observer did a better job of filtering out noise from the velocity estimate but resulted in a steady state error in the velocity estimate unless friction was included in the plant model. We then implemented a *Third-Order Observer* which included the friction as an additional state variable and showed how this eliminated the steady state error in the observer.

We then considered *Proportional*, *Derivative*, and *Integral Feedback* for controlling the wheel speed and position.

We also experimentally determined the *Coulomb* and *Viscous* components of the friction model.

CHAPTER 4

Stabilizing the Inverted Pendulum

In this chapter, we discuss the problem of stabilizing the pendulum in the inverted position. Linearizing Eq. (2.15) around $\theta = \pi$ gives

$$\ddot{\theta} - a\theta = -b_p(u - F)$$
$$\ddot{\theta}_r = b_r(u - F)$$

(4.1)

where θ denotes deviations from the value π. To begin with we will also neglect the friction torques. To stabilize the pendulum we can use the control law

$$u = -k_{pp}\theta - k_{dp}\dot{\theta}$$

(4.2)

which is a PD controller. Neglecting friction and inserting this control law in the Eq. (4.1) we find that the closed loop system is given by

$$\ddot{\theta} - b_p k_{pp}\dot{\theta} - (b_p k_{pp} + a)\theta = 0$$

Requiring that the closed loop system has the characteristic polynomial

$$A(s) = s^2 + 2\zeta\omega_0 s + \omega_0^2$$

we find that the controller coefficients are given by

$$k_{pp} = -\frac{\omega_0^2 + a}{b}$$

$$k_{dp} = -\frac{2\zeta\omega_0}{b}$$

(4.3)

Experiment 12 (Simple Stabilization). Try to stabilize the pendulum in the upright position with the control law (4.2). A reasonable parameter choice is $k_{pp} = 145$ and $k_{dp} = 16.4$. Observe the behavior of the system. You may have to support the pendulum manually to keep it stabilized. Explain what happens. *Note: Be sure that you start the controller each time with the pendulum at rest in the hanging position. After starting the controller, move the pendulum up to the inverted position manually. Explain why this is necessary to begin with the pendulum at rest in the downward position.*

To understand what happens in the experiment we will analyze the complete system. Assume that the pendulum initially has the angle θ_0 and that there is a disturbance torque d acting on the pendulum. The system is then described by the equations

$$\ddot{\theta} - b_p k_{dp} \dot{\theta} - (b_p k_{pp} + a)\theta = d \tag{4.4}$$

$$\ddot{\theta}_r + b_r k_{dp} \dot{\theta} + b_r k_{pp} \theta = 0 \tag{4.5}$$

where d represents disturbance torque on the pendulum. Assume that the pendulum is initially at rest at the angle θ_0 and that the control signal also is zero initially. Taking Laplace transforms and solving the equations we find the following expressions for the pendulum and rotor angles.

$$\Theta(s) = \frac{s - b_p k_{dp}}{s^2 - b_p k_{dp}s - a - b_p k_{pp}}\theta_0 + \frac{1}{s^2 - b_p k_{dp}s - a - b_p k_{pp}}D(s)$$

$$\tag{4.6}$$

$$\Theta_r(s) = -\frac{b_r(k_{dp}s + ak_{pp})(s - b_p k_{dp})}{s^2(s^2 - b_p k_{dp}s - a - b_p k_{pp})}\theta_0 - \frac{b_r b_p(k_{dp}s + k_{pp})}{s^2(s^2 - b_p k_{dp}s - a - b_p k_{pp})}D(s)$$

It follows from these equations that an initial offset in the angle in steady state the pendulum will result in the rotor having constant speed

$$\dot{\theta}_r = \frac{b_p b_r a k_{dp} k_{pp}}{a + b_p k_{pp}}\theta_0$$

It also follows that a constant disturbance torque d_0 in steady state will give a constant acceleration

$$\ddot{\theta}_r = \frac{b_r b_p k_{pp}}{a + b_p k_{pp}}d_0$$

A small disturbance torque from the cables will thus easily make the rotor velocity reach the saturation limit. This means that the PD controller will fail after a short time. To obtain a practical system it is therefore necessary to introduce feedback from the rotor velocity.

The Up-Down Transformation

It is difficult to make experiments with the inverted pendulum since the system is open loop unstable. If you make a slight error in programming or in choice of gains, the pendulum will fall down and the motion may be quite violent. Here, we will demonstrate a neat transformation that allows you to determine controllers to stabilize the inverted position but test them with the pendulum hanging down. Linearizing Eq. (2.15) around $\theta = 0$ gives

$$\ddot{\theta} + a\theta = -b_p(u - F)$$
$$\ddot{\theta}_r = b_r(u - F) \tag{4.7}$$

The key idea is based on the fact that the only difference in the model of the system is the sign of the coefficient a of the term θ in the system equation, compare Eqs. (4.7) and (4.1). We illustrate this with an example.

Assume that we design control laws so that the closed loop system has the characteristic polynomial

$$s^2 + 2\zeta\omega_0 s + \omega_0^2$$

The controller parameters are given by

$$k_{pp}^{down} = -\frac{\omega_0^2 - a}{b_p}$$

$$k_{dp}^{down} = -\frac{2\zeta\omega_0}{b_p}$$

when the pendulum is hanging down and

$$k_{pp}^{up} = -\frac{\omega_0^2 + a}{b} = k_{pp}^{down} - \frac{2a}{b_p}$$

$$k_{dp}^{up} = -\frac{2\zeta\omega_0}{b} = k_{dp}^{down}$$

when the pendulum is standing upright. In this particular case, the up–down transformation is simply to increase proportional gain by $2a/b$ and to keep the derivative gain.

4.1   CONTROLLABILITY

To avoid that the rotor velocity reaches the saturation limit it would be desirable to attempt to control both the pendulum angle and the rotor velocity. An even more ambitious scheme would be to also control the angle of the rotor. The first question we have to answer is if it is possible to do this with only one control variable. For that purpose we will investigate the controllability of the system. Introduce the state variables

$$x_1 = \theta, \quad x_2 = \dot{\theta}, \quad x_3 = \theta_r, \quad x_4 = \dot{\theta}_r$$

Neglecting the friction force, Eq. (4.7) can then be written as

$$\frac{dx}{dt} = \begin{pmatrix} 0 & 1 & 0 & 0 \\ a & 0 & 0 & 0 \\ 0 & 0 & 0 & 1 \\ 0 & 0 & 0 & 0 \end{pmatrix} x + \begin{pmatrix} 0 \\ -b_p \\ 0 \\ b_r \end{pmatrix} u = Ax + Bu \tag{4.8}$$

The controllability matrix is

$$W_c = \begin{pmatrix} B & AB & A^2 B & A^3 B \end{pmatrix} = \begin{pmatrix} 0 & -b_p & 0 & -ab_p \\ -b_p & 0 & -ab_p & 0 \\ 0 & b_r & 0 & 0 \\ b_r & 0 & 0 & 0 \end{pmatrix}$$

This matrix is full rank and the system is thus controllable. This means that the dynamics of the closed loop system can be shaped arbitrarily using only one control variable.

4.2   CONTROL OF THE PENDULUM AND THE WHEEL VELOCITY

Since the pendulum is influenced by the acceleration of the wheel, it may happen that the wheel velocity saturates after a while. It is thus desirable to try to achieve the dual goals of stabilizing the pendulum and to keep the wheel velocity small. To achieve this we will use the control law

$$u = -k_{pp}\theta - k_{dp}\dot{\theta} + k_{dr}(\omega_{ref} - \omega) \tag{4.9}$$

Inserting this control into Eq. (4.1) we find that the closed loop system is described by

$$\ddot{\theta} - b_p k_{dp}\dot{\theta} - (a + b_p k_{pp})\theta - b_p k_{dr}\omega = -b_p k_{dr}\omega_{ref} + d$$
$$b_r(k_{dp}\dot{\theta} + k_{pp}\theta) + \dot{\omega} + b_r k_{dr}\omega = b_r k_{dr}\omega_{ref} \tag{4.10}$$

The closed loop system has the characteristic polynomial

$$A(s) = s^3 + (-b_p k_{dp} + b_r k_{dr})s^2 - (a + b_p k_{pp})s + ab_r k_{dr}$$

The characteristic polynomial for a standard third-order system is

$$(s + \alpha\omega_0)(s^2 + 2\zeta\omega_0 s + \omega_0^2) = s^3 + (\alpha + 2\zeta)\omega_0 s^2 + (1 + 2\alpha\zeta)\omega_0^2 s + \alpha\omega_0^3$$

Identifying coefficients of equal powers of s in these polynomials we find that the controller parameters are given by

$$k_{pp} = -\frac{(1 + 2\alpha\zeta)\omega_0^2 + a}{b_p}$$

$$k_{dp} = -\frac{(\alpha + 2\zeta)\omega_0 a + \alpha\omega_0^3}{ab_p} \qquad (4.11)$$

$$k_{dr} = -\frac{\alpha\omega_0^3}{ab_r}$$

The controller parameters can be computed using the following Matlab program.

```
function [kpp,kdp,kdr]=pddcontrol(w0,zeta,alpha,a,bp,br)
%Computation of feedback gains for stabilization of pendulum
%and control rotor speed
kpp=-((1+2*alpha*zeta)*w0^2+a)/bp;
kdp=-((alpha+2*zeta)*w0*a+alpha*w0^3)/a/bp;
kdr=-alpha*w0^3/a/br
```

It remains to select suitable closed loop poles. Choosing $\omega_0 = 1.5\omega_p$, $\zeta = 0.707$ and $\alpha = 0.2$ gives $k_{pp} = -282$, $k_{dp} = -25.4$ and $k_{dr} = -0.0302$. If we instead choose $\omega_0 = 2*\omega_p$ we get instead $k_{pp} = -444$, $k_{dp} = -39.6$ and $k_{dr} = -0.0716$ and $\omega_0 = 2.5*\omega_p$ gives $k_{pp} = -654$, $k_{dp} = -59.8$ and $k_{dr} = -0.1405$.

> **Experiment 13 (Stabilization and Control of Wheel Velocity).** Try to stabilize the pendulum in the upright position with the control law (4.9). Find suitable values of the parameters k_{pp}, k_{dp}, and k_{dr}. Investigate the behavior of controller with different closed loop poles. Explore the system by tapping the pendulum with a ruler or a metal rod. Investigate the behavior of the pendulum and the wheel. Explain your findings theoretically.

Figures 4.1, 4.2, and 4.3 show results from experiments with these controllers. In the experiments a disturbance has been introduced by hitting the pendulum with a pencil. Three controllers with the parameters given previously have been investigated. The closed loop poles have the same pattern, $\zeta = 0.707$ and $\alpha = 0.2$, but the magnitudes of the poles are different, $\omega_0 = \omega_p$, $\omega_0 = 1.5\omega_p$, and $\omega_0 = 2\omega_p$.

The figures show that all controllers behave well. The pendulum angle returns quickly to the upright position. Notice that there is an offset in the pendulum in several of the experiments,

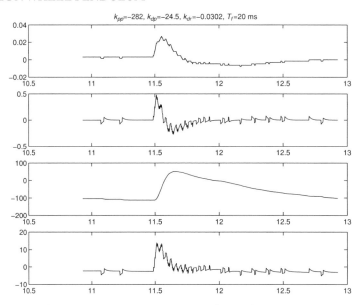

FIGURE 4.1: Results of experiment with stabilization of a pendulum using PD control of pendulum and control of wheel speed. The controller was designed to give $\omega_0 = \omega_p$, $\zeta = 0.707$, and $\alpha = 0.2$. The controller parameters are $k_{pp} = -282$, $k_{dp} = -24.5$, and $k_{dr} = -0.0302$.

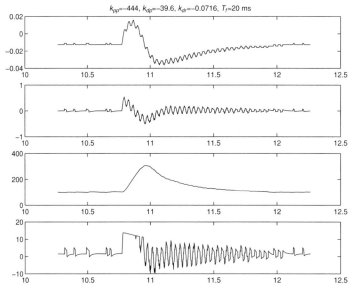

FIGURE 4.2: Results of experiment with stabilization of a pendulum using PD control of pendulum and control of wheel speed. The controller was designed to give $\omega_0 = 1.5\omega_p$, $\zeta = 0.707$, and $\alpha = 0.2$. The controller parameters are $k_{pp} = -444$, $k_{dp} = -39.6$, and $k_{dr} = -0.0716$.

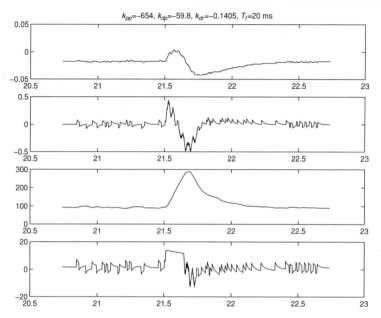

k_{pp}=−654, k_{dp}=−59.8, k_{dr}=−0.1405, T_f=20 ms

FIGURE 4.3: Results of experiment with stabilization of a pendulum using PD control of pendulum and control of wheel speed. The controller was designed to give $\omega_0 = 2.0\omega_p$, $\zeta = 0.707$, and $\alpha = 0.2$. The controller parameters are $k_{pp} = -654$, $k_{dp} = -59.8$, and $k_{dr} = -0.1405$.

indicating that the pendulum angle was not zero when the system was initialized. The pendulum is actually standing upright in spite of this as predicted by the theory. There are small steady state fluctuations around the steady state. The wheel spins at approximately constant speed after the disturbance. The wheel velocity is not zero because of disturbance torques from the cables and the offset in the angle measurement. The quantization of the angle sensor is clearly seen in the figures.

The closed loop system has an oscillatory mode with undamped natural frequency ω_0 and damping $\zeta = 0.707$ and a real mode $-\alpha\omega_0$ with $\alpha = 0.2$. The oscillatory mode decays at the rate $e^{-\zeta\omega_0 t}$ and the real mode decays at the rate $e^{-\alpha\omega_0 t}$. An inspection of the figures shows that the oscillatory mode dominates the response of the pendulum and that the real mode dominates the response of the wheel. It is thus possible to associate the different modes to the pendulum and the wheel. It is necessary to move the wheel in order to stabilize the pendulum, but it is also desired to keep the wheel speed limited to avoid saturation. It is, therefore, natural to make the wheel pole slower in the design. We did this by choosing a small value of α.

There is a slight difference between Figure 4.1 and Figures 4.2 and 4.3 because the real mode is larger in the last two figures. The reason for that is that the disturbance was larger and the controller hit the saturation limit in Figures 4.2 and 4.3. There is an oscillation in Figure 4.2

which is not noticeable in the other figures. This is due to a vibrational mode of the pendulum which was excited when the disturbance was introduced.

The trade-offs in the design can be judged from Figures 4.1, 4.2, and 4.3. The speed of recovery increases with higher bandwidths and higher controller gains but the fluctuations in the control signal also increases. The fluctuations in the control signal in Figure 4.1 are about 10% of the span of the control signal but they are about 25% in Figure 4.3.

4.3   LOCAL BEHAVIOR

We will now take a closer look of the stabilizing controller when there are no disturbances. Figures 4.4, 4.5, and 4.6 show the behavior of the system in three experiments using the same controllers as in the tapping experiment.

The figures show that all controllers are able to control the pendulum angle with a precision corresponding to the resolution of the angle sensor and that the behavior of the system is nonlinear. The pendulum error occasionally jumps with one resolution. These jumps cause large jumps in the estimated angular velocity of the pendulum and in turn jumps in the control signal. The jumps in the velocity are the same in all cases but the jumps in the control signal are larger for systems with larger values of the control gain k_{dp}. Also, notice that the angular velocity of the wheel does not have the jumps noticeable in the other signals.

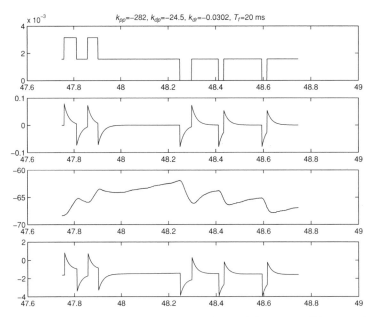

FIGURE 4.4: Results of experiment with stabilization of a pendulum using PD control of pendulum and control of wheel speed. The controller parameters are $k_{pp} = -282$, $k_{dp} = -24.5$, and $k_{dr} = -0.0302$.

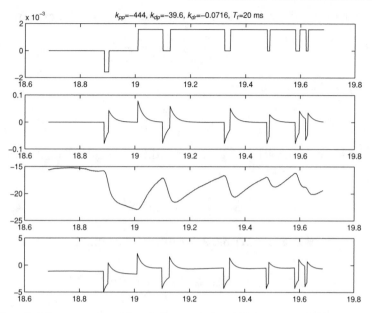

FIGURE 4.5: Results of experiment with stabilization of a pendulum using PD control of pendulum and control of wheel speed. The controller parameters are $k_{pp} = -444$, $k_{dp} = -39.6$, and $k_{dr} = -0.0716$.

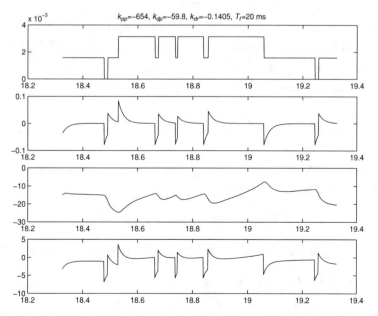

FIGURE 4.6: Results of experiment with stabilization of a pendulum using PD control of pendulum and control of wheel speed. The controller parameters are $k_{pp} = -654$, $k_{dp} = -59.8$, and $k_{dr} = -0.1405$.

A simple calculation shows the order of magnitude. The resolution of the angle sensor is

$$\Delta_\theta = \frac{2\pi}{4000} \approx 0.00157$$

When the velocity is filtered by the filter

$$G(s) = \frac{s}{1 + sT}$$

it follows that the jump in the velocity signal caused by an angle change of one increment of the encoder is

$$\Delta_{\dot\theta} = \frac{\Delta_\theta}{T} \approx 0.0785$$

The jumps in the signal $\dot\theta$ in all figures have this value. The jump in the control signal is then

$$\Delta_u = |k_{pp}|\Delta_\theta + |k_{dp}|\Delta_{\dot\theta} = \left(k_{pp} + \frac{k_{dp}}{T}\right)\Delta_\theta \qquad (4.12)$$

For the controller with the lowest gains we have $\Delta_u = 2.37$ which corresponds to 12% of the span of the control signal. For the other controllers the corresponding values are $\Delta_u = 3.81$ and $\Delta_u = 5.72$, or 19% and 29% of the span of the control signal. These values agree very well with the results of the experiments shown in Figures 4.4, 4.5, and 4.6.

In summary, we have found that all controllers keep the pendulum angle close to the upright position within the resolution of the angle sensor. The main difference is in the fluctuations in the control signal, which can be estimated by Eq. (4.12). Notice in particular the role of the filtering time constant T in (4.12).

4.4   STABILIZATION OF PENDULUM AND CONTROL OF WHEEL ANGLE

Next, we will attempt to keep tighter control of the wheel by having feedback from the wheel angle. The control strategy then becomes

$$u = k_{pp}(\theta_0 - \theta) - k_{dp}\dot\theta - k_{pr}\theta_r - k_{dr}\dot\theta_r \qquad (4.13)$$

Inserting this control into Eq. (4.1) we find that the closed loop system is described by

$$\begin{aligned} \ddot\theta - b_p k_{dp}\dot\theta + (a - b_p k_{pp})\theta - b_p(k_{dr}\dot\theta_r + k_{pr}\theta_r) &= -b_p k_{pr}\theta_{rref} + d \\ b_r(k_{dp}\dot\theta + k_{pp}\theta) + \ddot\theta_r + b_r k_{dr}\dot\theta_r + b_r k_{pr}\theta_r &= b_r k_{pr}\theta_{rref} \end{aligned} \qquad (4.14)$$

The closed loop system has the characteristic polynomial

$$A(s) = s^4 + (-b_p k_{dp} + b_r k_{dr})s^3 - (a + b_p k_{pp} + b_r k_{pr})s^2 - ab_r k_{dr}s - ab_r k_{pr}$$

The characteristic polynomial for a fourth-order system can be written as

$$(s^2 + 2\zeta_1\omega_1 s + \omega_1^2)(s^2 + 2\zeta_2\omega_2 s + \omega_2^2) = s^4 + 2(\zeta_1\omega_1 + \zeta_2\omega_2)s^3$$
$$+ (\omega_1^2 + 4\zeta_1\zeta_2\omega_1\omega_2 + \omega_2^2)s^2 + 2(\zeta_1\omega_2 + \zeta_2\omega_1)\omega_1\omega_2 s + \omega_1^2\omega_2^2$$

Identification of coefficients of equal powers of s with the characteristic polynomial gives the following equations.

$$-b_p k_{dp} + b_r k_{dr} = 2(\zeta_1\omega_1 + \zeta_2\omega_2)$$
$$-a - b_p k_{pp} + b_r k_{pr} = (\omega_1^2 + 4\zeta_1\zeta_2\omega_1\omega_2 + \omega_2^2)$$
$$-a b_r k_{dr} = 2(\zeta_1\omega_2 + \zeta_2\omega_1)\omega_1\omega_2$$
$$-a b_r k_{pr} = \omega_1^2\omega_2^2$$

Solving these equations with we find that the controller parameters are thus given by

$$k_{pp} = -\frac{a(\omega_1^2 + 4\zeta_1\zeta_2\omega_1\omega_2 + \omega_2^2) + \omega_1^2\omega_2^2 + \omega_p^4}{ab_p}$$

$$k_{dp} = -\frac{2a(\zeta_1\omega_1 + \zeta_2\omega_2) + 2(\zeta_1\omega_2 + \zeta_2\omega_1)\omega_1\omega_2}{ab_p}$$

$$k_{pr} = -\frac{\omega_1^2\omega_2^2}{ab_r}$$

$$k_{dr} = -\frac{2(\zeta_1\omega_2 + \zeta_2\omega_1)\omega_1\omega_2}{ab_r}$$

(4.15)

gives a closed loop system with the characteristic polynomial

$$(s^2 + 2\zeta_1\omega_1 s + \omega_1^2)(s^2 + 2\zeta_2\omega_2 s + \omega_2^2) = s^4 + 2(\zeta_1\omega_1 + \zeta_2\omega_2)s^3$$
$$+ (\omega_1^2 + 4\zeta_1\zeta_2\omega_1\omega_2 + \omega_2^2)s^2 + 2(\zeta_1\omega_2 + \zeta_2\omega_1)\omega_1\omega_2 s + \omega_1^2\omega_2^2$$

The controller parameters can be computed with the following Matlab function

```
function [kpp,kdp,kpr,kdr]=pdpdcontrol(w1,zeta1,w2,zeta2,a,bp,br)
%Computation of feedback gains for stabilization of pendulum
%and control wheel angle
kpp=-(a*(w1^2+4*zeta1*zeta2*w1*w2+w2^2)+w1^2*w2^2+a^2)/a/bp;
kdp=-(2*a*(zeta1*w1+zeta2*w2)+2*(zeta1*w2+zeta2*w1)*w1*w2)/a/bp;
kpr=-w1^2*w2^2/a/br;
kdr=-2*(zeta1*w1+zeta2*w1)*w1*w2/a/br;
```

Choosing $\omega_1 = \omega_2 = \omega_p$, $\zeta_1 = 0.707$, and $\zeta_2 = 1$ gives $k_{pp} = -496$, $k_{dp} = -56.0$, $k_{pr} = -0.396$, and $k_{dr} = -0.153$. With $\omega_1 = \omega_p$, $\omega_2 = 0.5\omega_p$, $\zeta_1 = 1$, and $\zeta_2 = 0.707$ we get $k_{pp} = -284$, $k_{dp} = -32.0$, $k_{pr} = -0.0990$, and $k_{dr} = -0.0763$.

Experiment 14 (Stabilization of Pendulum and Control of Wheel Angle). Choose reasonable parameters for the controller (4.13) based on the calculations above. Run the controller and observe what happens when the reference value is changed. Also, investigate what happens when a disturbance torque is applied to the pendulum.

Taking Laplace transform of Eq. (4.14) and solving for the transforms of the angles we find

$$\Theta(s) = -\frac{b_p k_{pr} s^2}{A(s)}\Theta_{rref}(s) + \frac{s^2 + b_r k_{dr} s + b_r k_{pr}}{A(s)} D(s)$$

$$\Theta_r(s) = \frac{b_r k_{pr}(s^2 - a)}{A(s)}\Theta_{rref}(s) - \frac{b_r(k_{dp} s + k_{pp})}{A(s)} D(s) \qquad (4.16)$$

$$A(s) = s^4 + (-b_p k_{dp} + b_r k_{dr})s^3 + (-b_p k_{pp} + b_r k_{pr} - a)s^2 - ab_r k_{dr} s - ab_r k_{pr}$$

4.5 EFFECT OF OFFSET IN ANGLE AND DISTURBANCE TORQUES

It is easy to initialize the angle sensor so that it gives the right signal when the pendulum hangs down. It is more difficult to get the correct calibration when the pendulum stands up. If the pendulum is calibrated in the down position and the up position is calculated there may be calibration errors. We will now investigate the effects of such errors. At the same time we will also investigate the effect of a disturbance torque on the pendulum. We will start with an experiment.

Experiment 15 (Effect of a Disturbance Torque). Stabilize the pendulum in the upright position with the controller (4.9). Apply a small disturbance torque to the pendulum, e.g., by using a rubber band or an elastic rod. Observe the behavior of the system.

Assume that the control law is

$$u = k_{pp}(\delta - \theta) - k_{dp}\dot{\theta} - k_{pr}\theta_r - k_{dr}\dot{\theta}_r$$

where δ is a constant offset. Notice that we have included a term proportional to wheel angle. We can obtain the results for feedback with wheel rate only by setting $k_{pr} = 0$. Inserting this control law into Eq. (4.1) we get

$$\ddot{\theta} - b_p k_{dp}\dot{\theta} - (a + b_p k_{pp})\theta - b_p(k_{dr}\dot{\theta}_r + k_{pr}\theta_r) = -b_p k_{pp}\theta_0 + b_p d$$
$$b_r(k_{dp}\dot{\theta} + k_{pp}\theta) + \ddot{\theta}_r + b_r k_{dr}\dot{\theta}_r + b_r k_{pr}\theta_r = b_r k_{pp}\theta_0$$

The closed loop system has the characteristic polynomial

$$A(s) = s^4 + (-b_p k_{dp} + b_r k_{dr})s^3 + (-a - b_p k_{pp} + b_r k_{pr})s^2 - ab_r k_{dr}s - ab_r k_{pr}$$

Taking Laplace transforms and solving the equations we get

$$\Theta(s) = -\frac{b_p k_{pp}s}{A(s)}\delta + \frac{b_p(s^2 + b_r k_{dr}s + b_r k_{pr})}{A(s)}D(s)$$
$$\Theta_r(s) = \frac{b_r k_{pp}(s^2 - a)}{s\,A(s)}\delta - \frac{b_r b_p(k_{dp}s + k_{pp})}{A(s)}D(s)$$

(4.17)

A calibration error δ in the pendulum angle gives no steady state error in the pendulum angle but it gives a steady state drift of the wheel with rate $\delta k_{pp}/k_{pr}$. A constant disturbance torque d_0 gives a steady state pendulum error $b_p d_0/a$ and a drift of the wheel with rate $b_p k_{pp}d_0/ak_{pr}$.

4.6 FEEDBACK OF WHEEL RATE ONLY

We can obtain the effects of calibration error and disturbance torque on the pendulum for the case of wheel rate feedback only by setting $k_{pr} = 0$ in Eq. (4.16). This gives

$$\Theta(s) = -\frac{b_p k_{pp}}{A(s)}\delta + \frac{b_p(s + b_r k_{dr})}{A(s)}D(s)$$
$$\Theta_r(s) = \frac{b_r k_{pp}(s^2 - a)}{s^2 A(s)}\delta - \frac{b_r b_p(k_{dp}s + k_{pp})}{s\,A(s)}D(s)$$

(4.18)

$$A(s) = s^3 + (-b_p k_{dp} + b_r k_{dr})s^2 - (b_p k_{pp} + a)s - ab_r k_{dr}$$

A calibration error δ in the pendulum angle gives non steady state error in the pendulum angle but a steady state drift of the wheel with rate $\delta k_{pp}/k_{dr}$. A constant disturbance torque d_0 also gives a steady state error $-b_p d_0/a$ in the pendulum angle and a steady state drift of the wheel with rate $b_p k_{pp}d_0/ak_{dr}$. Notice that a positive torque gives a negative steady state error in the pendulum angle. This explains the counterintuitive behavior seen in Experiment 15 where the pendulum moves in the direction opposite to the applied torque.

4.7   A REMARK ON CONTROLLER DESIGN

We have used pole placement to design the controllers. This means that we have specified the closed loop poles and computed the controller parameters that give the desired closed loop poles. The particular technique we used was straightforward. The control law was combined with the process model and we derived the closed loop characteristic polynomial and matched it with the desired characteristic polynomial. This method is simple and direct and it gives analytic expressions for the controller gains. This means that it is easy to see how controller gains are influenced by system parameters and specifications. The method is very easy to use for systems of low-order like the pendulum. An alternative which is particularly useful for more complex systems is to use state space methods. There are algorithms in Matlab that perform compute the state feedback directly from a state model. We illustrate this by an example.

Example [Computing Feedback Gains Using Matlab]

A state model for the system is given by Eq. (4.8). The feedback gains that gives a closed loop system with the characteristic equation

$$(s^2 + 1.414\omega_0 s + \omega_0^2)((s^2 + 2\omega_0 s + \omega_0^2)$$

where $\omega_0 = \omega_p = 8.8557$ can be computed by the following Matlab program.

```
%State space computation of controller gains
systpar %Get system parameters
A=[0 1 0 0;a 0 0 0;0 0 0 1;0 0 0 0]
B=[0;-bp;0;br]
P=roots([1 2*zeta1*w1 w1^2]);
P=[P;roots([1 2*zeta2*w2 w2^2])]
K=acker(A,B,P)
K=place(A,B,P)
```

Running this program we get the following result

```
Warning: Pole locations are more than 10% in error.
K = -495.5568   -55.9593    -0.3961    -0.1527
??? Error using ==> place
Can't place poles with multiplicity greater than rank(B).
```

The error message indicates that numerical difficulties are encountered in the program `acker` and that the program `place` is unable to solve the problem. The results can be compared with the results of the direct calculation of the controller parameters by the program.

```
function [kpp,kdp,kpr,kdr]=pdpdcontrol(w1,zeta1,w2,zeta2,a,bp,br)
%Computation of feedback gains for stabilization of pendulum
%and control rotor angle
kpp=-(a*(w1^2+4*zeta1*zeta2*w1*w2+w2^2)+w1^2*w2^2+a^2)/a/bp;
kdp=-(2*a*(zeta1*w1+zeta2*w2)+2*(zeta1*w2+zeta2*w1)*w1*w2)/a/bp;
kpr=-w1^2*w2^2/a/br;
kdr=-2*(zeta1*w1+zeta2*w1)*w1*w2/a/br;
```

This program gives

```
K = -495.5568   -55.9593   -0.3961   -0.1527
```

The parameters are identical with the results obtained using acker. An analysis of Eq. (4.15) shows that there are no real numerical difficulties in computing the gains. The method used in acker is, however, inherently poorly conditioned. Considering the precision in the model parameters it is not reasonable to give controller parameters with more than three significant digits.

Summary

In this chapter, we investigated the control of the pendulum and wheel together. For the problem of stabilizing the pendulum in the inverted position, we linearized the nonlinear equations about $\theta = \pi$ and considered the resulting fourth-order linear system.

We saw that feedback of the pendulum states is not sufficient to control the system because the wheel velocity "runs away." Incorporating feedback of the wheel velocity, it is possible to stabilize the pendulum and simultaneously regulate the wheel velocity.

We then considered the problem of controlling all four states. We showed that the fourth-order system is controllable from the single input available and we considered the effect of disturbances on the system response.

The controllers in this chapter are *local* in the sense that the initial conditions must be sufficiently close to the equilibrium point for the closed loop system to be stable. For initial conditions too far away from the equilibrium, or for large disturbances which move the system too far away from the equilibrium the pendulum will fall down. In the next chapter, we consider the problem of starting the pendulum and wheel in arbitrary initial conditions. This problem requires consideration of nonlinear control theory.

CHAPTER 5

Swinging Up the Pendulum

5.1 NONLINEAR CONTROL

So far we have discussed the identification of the parameters needed in the dynamic description of the Reaction Wheel Pendulum, we have discussed stabilization of the motor position and velocity, and stabilization of the pendulum angle in both the downward and the inverted positions using the linear approximation of the nonlinear dynamics. In this chapter, we will introduce some ideas from nonlinear control theory and show how these ideas may be used to control large angular movements of the pendulum. As an illustration we will discuss the interesting problem of swingup control. The swingup problem is to move the pendulum from the downward hanging position up into the inverted position. Since the angle of the pendulum must change by 180°, the assumption that $\sin(\theta)$ can be well approximated by θ is no longer a good one and we must deal with the nonlinear equations of motion directly. In the next chapter, we will combine the swingup control from this chapter with the stabilization control from the previous chapter as an illustration of so-called *Switching Control* to design a controller that swings the pendulum up and catches it.

5.2 SOME BACKGROUND ON NONLINEAR CONTROL

This section provides some background on nonlinear control theory for completeness. The reader already familiar with these notions may safely skip this section and move directly to Section 5.3.

5.2.1 State Space, Equilibrium Points, and Stability

The concepts of transfer functions, poles and zeros, and frequency response do not extend to systems described by nonlinear differential equations. We must, therefore, look elsewhere for concepts and techniques that we can use for analysis and design. In this section, we will discuss the notion of *Passivity* which, as we shall see, is related to energy dissipation and provide us with an elegant and powerful method for designing nonlinear controllers to control the Reaction Wheel Pendulum, particularly for the swingup control problem.

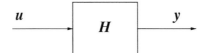

FIGURE 5.1: Input/output system.

Let a system be described by an input/output relationship $H(u, y)$ as shown in Figure 5.1. The function H is just meant to represent the "rule" that the system uses to produce an output y in response to an input u. For a linear system, of course, H may be represented as a transfer function giving the ratio of the Laplace transforms of the input and output. For a nonlinear system, one may think of the "operator" H as a shorthand way to denote a set of differential equations

$$\dot{x} = f(x) + g(x)u \qquad (5.1)$$
$$y = h(x) \qquad (5.2)$$

relating the input and output, where $x \in \Re^n$ is the system state vector. The functions $f(x)$ and $g(x)$ are vector fields on $\Re^n$ and $h(x)$ is the output function.

Definition 5.1. *An Equilibrium or Fixed Point of a system*

$$\dot{x} = f(x) \qquad (5.3)$$

is a vector $\bar{x}$ such that $f(\bar{x}) = 0$.

If the initial condition for the system (5.3) satisfies $x(t_0) = \bar{x}$ then the function $x(t) \equiv \bar{x}$ for $t > t_0$ can be seen to be a solution of (5.3). In other words, if the system (5.3) starts initially at the equilibrium, then it remains at the equilibrium thereafter.

Example 5.1. The simple pendulum equation

$$\ddot{\theta} + \sin \theta = 0 \qquad (5.4)$$

is written in the state space form (5.3) by setting

$$x = \begin{pmatrix} x_1 \\ x_2 \end{pmatrix} = \begin{pmatrix} \theta \\ \dot{\theta} \end{pmatrix}$$

Then the second-order system (5.4) is equivalent to the two first-order equations

$$\dot{x}_1 = x_2$$
$$\dot{x}_2 = -\sin x_1$$

and the vector field $f(x)$ is given by

$$f(x) = \begin{pmatrix} x_2 \\ -\sin x_1 \end{pmatrix}$$

Equating $f(x)$ to zero to find the equilibrium points leads to the condition that

$$x_2 = 0, \quad x_1 = n\pi, \quad n = 0, 1, 2, \ldots$$

Thus the equilibrium solutions of the simple pendulum equation correspond to initial conditions where the velocity is zero and the pendulum is straight down ($n = $ even) or straight up ($n = $ odd). This corresponds to our intuitive notion of an equilibrium configuration for the simple pendulum.

The question of stability deals with the solutions of the system for initial conditions away from the equilibrium. Intuitively, the equilibrium should be called stable if, for initial conditions close to the equilibrium, the solution remains close thereafter, as in the vertically downward equilibrium of the pendulum, and unstable if nearby solutions diverge from the equilibrium, as in the vertically upward equilibrium of the pendulum. We can formalize this notion into the following.

Definition 5.2. *The equilibrium solution $x(t) = \bar{x}$ is said to be*

i) *stable if and only if, for any $\epsilon > 0$ there exist $\delta(\epsilon) > 0$ such that*

$$\|x(t_0)\| < \delta \text{ implies } \|x(t)\| < \epsilon \text{ for all } t > t_0 \tag{5.5}$$

ii) *asymptotically stable if and only if it is stable and there exists $\delta > 0$ such that*

$$\|x(t_0)\| < \delta \text{ implies } \|x(t)\| \to 0 \text{ as } t \to \infty \tag{5.6}$$

iii) *exponentially stable if and only if there exist constants $k > 0$, $\gamma > 0$ such that*

$$x(t) < k\|x(t_0)\|e^{-\gamma(t-t_0)} \quad \text{for} \quad t \geq t_0$$

iv) *unstable if it is not stable*

The situation is illustrated by Figure 5.2 and says that the system is stable if the solution remains within a ball of radius ϵ around the equilibrium, so long as the initial condition lies in a ball of radius δ around the equilibrium. Notice that the required δ will depend on the given ϵ. To put it another way, a system is stable if "small" perturbations in the initial conditions, results in "small" perturbations from the equilibrium solution. For a stable system, the trajectory will remain within the ball of radius δ for all time. For an asymptotically (or exponentially) stable system, the trajectory will, in addition, return to the equilibrium point as $t \to \infty$, while for an

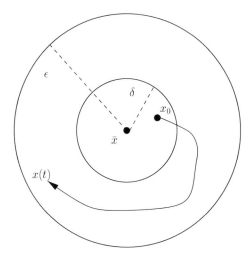

FIGURE 5.2: Illustrating the definition of stability.

unstable system, no matter how small δ is chosen, the trajectory will eventually leave the ball of radius ϵ. The above notions of stability are local in nature, that is, they may hold for initial conditions "sufficiently near" the equilibrium point but may fail for initial conditions farther away from the equilibrium. Stability (respectively, asymptotic stability) is said to be *global* if it holds for arbitrary initial conditions. Note that global stability can never hold in the quite common case that the nonlinear system has multiple equilibrium points.

5.2.2   Linearization of Nonlinear Systems

We know that a linear system

$$\dot{x} = Ax \tag{5.7}$$

is globally exponentially stable provided that all eigenvalues of the matrix A lie in the open left half of the complex plane and unstable if any eigenvalue of A has positive real part. For nonlinear systems, local stability can often be determined from the linear approximation of the nonlinear equations about the equilibrium point, in other words, by examining the system

$$\delta\dot{x} = A_{\bar{x}}\delta x \quad \text{where } \delta x = x - \bar{x} \text{ and } A_{\bar{x}} = \frac{\partial f}{\partial x}\Big|_{x=\bar{x}} \tag{5.8}$$

Theorem 5.1.   *Consider the nonlinear system (5.3) with $f(\bar{x}) = 0$. Let $A_{\bar{x}} = \frac{\partial f}{\partial x}\big|_{x=\bar{x}}$. Then*

i)   *the matrix $A_{\bar{x}}$ has all its eigenvalues in the open left half plane if and only if the equilibrium $\bar{x}$ of the nonlinear system (5.3) is locally exponentially stable*

ii)   *if the matrix $A_{\bar{x}}$ has one or more eigenvalues with positive real part, the equilibrium $\bar{x}$ of the nonlinear system (5.3) is unstable*

iii)   *if the matrix $A_{\bar{x}}$ has no eigenvalue with positive real part but one or more eigenvalues on the $j\omega$-axis, then stability of the equilibrium of the nonlinear system, (5.3), cannot be determined from the linear approximation*

In the latter case, we cannot say whether the equilibrium of the nonlinear system is stable, asymptotically stable, or unstable. We can say however, it cannot be exponentially stable.

Performing the indicated calculations for the simple pendulum equation (5.4) yields

$$A_0 = \begin{bmatrix} 0 & 1 \\ -1 & 0 \end{bmatrix} \qquad A_\pi = \begin{bmatrix} 0 & 1 \\ 1 & 0 \end{bmatrix}$$

with respect to the two equilibrium configurations $(0, 0)$ and $(0, \pi)$, respectively. The matrix A_π has eigenvalues at ± 1 which confirms our intuition that the inverted equilibrium is unstable. However, since the matrix A_0 has both eigenvalues on the imaginary axis at $\pm j$, we cannot conclude that the downward equilibrium is stable by examining only the linear approximation of the nonlinear equations. We will next state, but not prove two fundamental results, due to Lyapunov and LaSalle, respectively, that are among the most useful results for stability analysis and control design for nonlinear systems. The interested reader should refer to the material on *Lyapunov Stability Theory* and *LaSalle's Invariance Principle* in [8] for the complete details. We first need some additional background on the notions of Positive Definite Functions and Invariant Sets for nonlinear systems.

Definition 5.3.   *A scalar function $V : \Re^n \to \Re$ is said to be*

i)   Positive Semi-Definite *if and only if $V(0) = 0$ and $V(x) \geq 0$ for $x \neq 0$*

ii)   Positive Definite *if and only if $V(0) = 0$ and $V(x) > 0$ for $x \neq 0$*

iii)   Negative (Semi)-Definite *if and only if $-V(x)$ is Positive (Semi)-Definite*

As in all such notions one may attach the adjective local or global to the above definitions as the case may be. Locally, the level surfaces of a positive definite function V, given as solutions of $V(x) = C$, where C is a positive constant, are ellipsoids in $\Re^n$ (see Figure 5.3).

A positive definite function is like a norm. In fact, given the usual norm $\|x\|$ on R^n, the function $V(x) = x^T x = \|x\|^2$ is positive definite. More generally, given a symmetric matrix $P = (p_{ij})$ the scalar function

$$V(x) = x^T P x = \sum_{i,j=1}^n p_{ij} x_i x_i$$

is positive definite if and only if the matrix P has all eigenvalues positive.

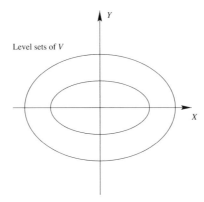

FIGURE 5.3: Level sets of a positive definite function.

Definition 5.4.

i) *Let $V(x) : \Re^n \to \Re$ be a continuous, positive definite, function with continuous first partial derivatives in a neighborhood of the origin in $\Re^n$. Then V is called a* Lyapunov Function Candidate *(for the system (5.3)).*

ii) *By the derivative of V along trajectories of (5.3), or the derivative of V in the direction of the vector field defining (5.3), we mean*

$$\dot{V}(t) = \nabla V^T f(x) = \frac{\partial V}{\partial x_1} f_1(x) + \cdots + \frac{\partial V}{\partial x_n} f_n(x). \tag{5.9}$$

Suppose that we evaluate the Lyapunov function candidate V at points along a solution trajectory $x(t)$ of (5.3) and find that $V(t)$ is decreasing for increasing t. Intuitively, since V acts like a norm, this must mean that the given solution trajectory must be converging toward the origin. This is the idea of Lyapunov stability theory.

Theorem 5.2. *The equilibrium solution of (5.3) is stable if there exists a Lyapunov function candidate V such that $\dot{V}$ is negative semi-definite along solution trajectories of (5.3), that is, if*

$$\dot{V} = \nabla V^T f(x) \leq 0. \tag{5.10}$$

Equation (5.10) says that the derivative of V computed along solutions of (5.3) is non-positive, which says that V itself is nonincreasing along solutions. Since V is a measure of how far the solution is from the origin, Eq. (5.10) says that the solution must remain near the origin. If a Lyapunov function candidate V can be found satisfying (5.10) then V is called a *Lyapunov Function* for the system (5.3). Note that Theorem 5.2 gives only a sufficient condition for stability of (5.3). If one is unable to find a Lyapunov function satisfying (5.10) it does not mean that the system is unstable. However, an easy sufficient condition for instability of (5.3) is for there to exist a Lyapunov function candidate V such that $\dot{V} > 0$ along at least one solution of the system.

Theorem 5.3. *The equilibrium of (5.3) is asymptotically stable if there exists a Lyapunov function candidate V such that $\dot{V}$ is strictly negative definite along solutions of (5.3), that is,*

$$\dot{V}(x) < 0. \tag{5.11}$$

The inequality (5.11) means that V is actually decreasing along solution trajectories of (5.3) and hence the trajectories must be converging to the equilibrium point.

The strict inequality in (5.11) may be difficult to obtain for a given system and Lyapunov function candidate. We, therefore, discuss LaSalle's Invariance Principle which can be used to prove asymptotic stability even when V is only negative semi-definite.

Definition 5.5. *A subset $\mathcal{D}$ of $\mathfrak{R}^n$ is said to be (Positively) Invariant for the system (5.3) if*

$$x(t_0) \in \mathcal{D} \implies x(t) \in \mathcal{D} \text{ for } t > t_0$$

Sets consisting of equilibrium points are clearly invariant according to the above definition as are trajectories themselves.

Theorem 5.4. *[8] Let $\mathcal{D} \subset \mathfrak{R}^n$ be a compact set that is positively invariant for the system (5.3). Let V be a continuously differentiable function such that $\dot{V} \leq 0$ in $\mathcal{D}$. Let Ω be the set of points in $\mathcal{D}$ where $\dot{V} = 0$. Let M be the largest invariant set in Ω. Then every solution starting in $\mathcal{D}$ approaches M as $t \to \infty$.*

The above theorem is known as *LaSalle's Theorem* and will prove extremely useful in our subsequent analysis. To illustrate the above concepts, we will consider again the simple pendulum example

$$\ddot{\theta} + \sin\theta = 0 \tag{5.12}$$

and investigate the stability properties of the equilibrium $x_1 = \theta = 0$, $x_2 = \dot{\theta} = 0$. We recall that the analysis based on the linear approximation of the nonlinear equations was inconclusive even though we know intuitively that this equilibrium is stable. Indeed, we know that all solutions that start near this equilibrium are periodic. Furthermore, we know that in practice, any amount of friction will cause the oscillations to eventually decay to zero so that, in this case, the equilibrium is asymptotically stable.

To confirm our intuition, we let E be the total energy of the pendulum, i.e.,

$$E = \frac{1}{2}\dot{\theta}^2 + 1 - \cos\theta$$

It is easy to see that E is positive definite near the downward equilibrium since the cosine function is less than one in absolute value and hence qualifies as a Lyapunov function candidate. Computing $\dot{E}$ along trajectories of (5.12) yields

$$\dot{E} = \dot{\theta}\ddot{\theta} + \sin\theta\dot{\theta}$$
$$= 0$$

where the latter equality follows by substituting $\ddot{\theta}$ from (5.12). Thus we have shown that the equilibrium is stable. If we include viscous friction in the model

$$\ddot{\theta} + b\dot{\theta} + \sin\theta = 0$$

then the same calculation gives us

$$\dot{E} = -b\dot{\theta}^2$$

However, we can still conclude only stability and not asymptotic stability because $\dot{E}$ is still only negative semi-definite (because $\dot{E}$ is zero when $\dot{\theta}$ is zero for any value of θ). We cannot rule out the possibility that the pendulum will "become stuck" at a configuration where the velocity is zero but the angle is not zero. Using LaSalle's Theorem, we compute the set $\dot{E} = 0$, which gives us $\dot{\theta} = 0$. Suppose then that a trajectory satisfies $\dot{\theta}(t) = 0$ for all t. This implies that $\ddot{\theta}(t) = 0$, whence, from (5.13), we must have $\sin(\theta) = 0$ and we conclude from LaSalle's Theorem that the largest invariant set contained in the set where $\dot{V} = 0$ consists only of the two equilibrium points. Therefore the downward configuration of the pendulum is (locally) asymptotically stable.

5.2.3   Passivity
We now wish to investigate the Input/Output system (5.1). To do this it is useful to introduce the notion of *Passivity*.

Definition 5.6. *We shall say that the system H described by (5.1) is* Passive *if there exists a (locally) positive definite scalar function S, called a* Storage Function, *such that*

$$\dot{S} \leq u^T y \rightarrow S(T) - S(0) \leq \int_0^T u^T(s)y(s)ds \tag{5.13}$$

For example, in a circuit made up of passive elements (resistors, capacitors, and inductors), the (effort) variable u represents current and the (flow) variable y represents voltage. The product $u^T y$ is thus the instantaneous power and the integral is the energy. Equation (5.13) says, therefore, that in a passive system the change in system energy is not greater than that supplied by the input.

Since we assume that the storage function is positive definite[1] it follows that a passive system is open loop stable since S qualifies as a Lyapunov function with $u = 0$.

Example 5.2.   Consider the mass–spring–damper system

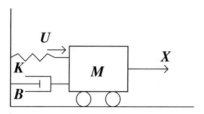

with equation of motion

$$M\ddot{x} + B\dot{x} + Kx = u$$

The total energy of the system is

$$H = \frac{1}{2}M\dot{x}^2 + \frac{1}{2}Kx^2$$

Then $\dot{H}$ satisfies

$$\begin{aligned}\dot{H} &= M\dot{x}\ddot{x} + Kx\dot{x} \\ &= \dot{x}(M\ddot{x} + Kx) \\ &= \dot{x}u - B\dot{x}^2 \leq \dot{x}u\end{aligned}$$

Therefore, the mass–spring–damper system is passive if we take as input the force u on the mass and as output y the velocity of the mass $v = \dot{x}$, but it is not passive if we take the position $y = x$ as output.

In fact, it can be shown that a linear system must have relative degree ≤ 1 in order to be passive. Since the transfer function from u to x, i.e.,

$$G_1(s) = \frac{X(s)}{U(s)} = \frac{1}{Ms^2 + Bs + K}$$

has relative degree two, it cannot be passive.

---

[1]A more complete treatment of passivity would assume that the storage function S is only positive semi-definite rather than positive definite. See [8] for details.

An advantage of passive systems is that they can be easily stabilized by proportional feedback of the passive output.

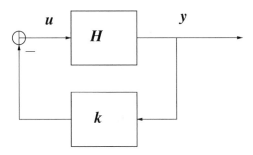

Setting $u = -ky$ we see that the Storage Function S satisfies

$$\dot{S} \leq u^T y = -ky^T y \leq 0$$

for any value of the gain k. This implies that S will continually decrease as long as the output is not zero. LaSalle's Theorem can then be used to investigate the asymptotic behavior of the system, as determined by the particular output function.

The above "infinite gain margin" is an important feature of passive systems. There are other properties of passive systems that can be exploited in control systems design, such as the fact that Parallel and Feedback interconnections of Passive (PR) systems are Passive (see Figures 5.4 and 5.5).

5.3   SWINGUP CONTROL OF THE REACTION WHEEL PENDULUM

Now, we return to the consideration of the Reaction Wheel Pendulum with dynamics given, as before, by

$$J\ddot{\theta} + mg\ell \sin(\theta) = -ku$$
$$J_r \ddot{\theta}_r = ku$$

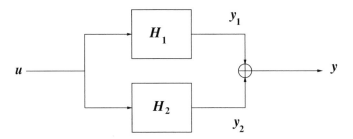

FIGURE 5.4: Parallel interconnection.

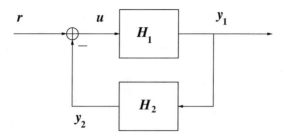

FIGURE 5.5: Feedback interconnection.

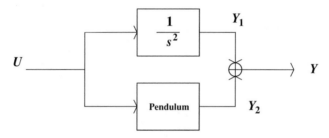

FIGURE 5.6: Reaction wheel system as a parallel interconnection.

We have noted that this system can be thought of as a parallel interconnection of the pendulum subsystem and the wheel subsystem (double integrator) with input u. If we write this system as shown in Figure 5.6 then the immediate task at hand is to define output functions y_1 and y_2 in such a way that this parallel interconnection is passive.

For the double integrator subsystem we can choose as output $y_1 = \dot{\theta}_r$ and storage function $S_1 = \frac{1}{2}\dot{\theta}_r^2$. We note that the disk subsystem cannot be passive from input u to output $\dot{\theta}_r$ since the double integrator system has relative degree two.

How shall we define y_2 so that the pendulum subsystem is passive? Let us return to the energy equation for the pendulum

$$E = \frac{1}{2}J\dot{\theta}^2 + mg\ell(1 - \cos(\theta)) \tag{5.14}$$

A similar calculation as before shows that

$$\dot{E} = -ku\dot{\theta} \tag{5.15}$$

We thus recover the previous result that the energy is constant along trajectories if $u = 0$ and show, in addition, that the pendulum subsystem is passive from $-ku$ to $\dot{\theta}$ with the energy E as storage function.

However, instead of taking the energy E as the storage function, we shall take

$$S_2 = \frac{1}{2}(E - E_{ref})^2$$

where E_{ref} represents a (constant) reference energy. Then

$$\dot{S}_2 = (E - E_{ref})\dot{E} = -k\dot{\theta}(E - E_{ref})u \tag{5.16}$$
$$= y_2 u \tag{5.17}$$

where we have taken as the output function $y_2 = -k\dot{\theta}(E - E_{ref})$. The reason for this choice of output y_2 and storage function S_2 will become evident shortly. It follows that the parallel interconnection is passive with output

$$y = k_v y_1 + k_e y_2$$

and storage function

$$S = k_v S_1 + k_e S_2$$
$$= \frac{1}{2}k_v \dot{\theta}_r^2 + \frac{1}{2}k_e(E - E_{ref})^2$$

We have chosen a linear combination of the storage function S_1 and S_2 with weights k_v and k_e, respectively, to allow extra design freedom in the controller. As we shall see, the constants k_v and k_e play the role of adjustable gains that can be designed to influence the transient response of the system.

Computing $\dot{S}$ along trajectories of the system yields

$$\dot{S} = (k_v \dot{\theta}_r - k_e k(E - E_{ref})\dot{\theta})u = yu$$

We then choose the control input u as

$$u = -k_u y = -k_u(k_v \dot{\theta}_r - k_e k(E - E_{ref})\dot{\theta}) \tag{5.18}$$

and we have

$$\dot{S} = -k_u y^2 \leq 0 \tag{5.19}$$

Therefore, the system is stable and LaSalle's Invariance Principle can now be used to determine the asymptotic behavior of the system. Setting $y_2 \equiv 0$ yields

$$u = k_u(k_e k(E - E_{ref})\dot{\theta} - k_v \dot{\theta}_r) \equiv 0 \tag{5.20}$$

It follows that the derivative $\dot{u} \equiv 0$ from which we get

$$k_u(k_e k\dot{E}\dot{\theta} + k_e k(E - E_{ref})\ddot{\theta} - k_v \ddot{\theta}_r) = 0$$

Since $u = 0$, $\ddot{\theta}_r = ku$, $\dot{E} = -ku\dot{\theta}$, and $\ddot{\theta} = -\frac{mgl}{J}\sin\theta - ku$, all of this reduces to

$$(E - E_{ref})\sin\theta = 0$$

This equation is very important and says that the closed loop system trajectories will converge to either $E = E_{ref}$ or $\sin\theta = 0$.

Therefore, for all initial conditions away from the equilibrium points, it follows that $E - E_{ref} \to 0$. In addition, it follows from (5.20) that $\dot{\theta}_r = 0$.

Remark 5.1 (The Effect of Saturation). An extremely important advantage of the Passivity-Based Control approach to the problem of swingup is that the analysis remains unchanged if, instead of using the control law (5.18), we use the saturated control

$$u = -\mathrm{sat}(k_u y) = -\mathrm{sat}(k_u(k_v\dot{\theta}_r - k_e k(E - E_{ref})\dot{\theta})) \qquad (5.21)$$

where $\mathrm{sat}(\cdot)$ denotes the saturation function. The Storage Function S, given by (5.18) then satisfies

$$\dot{S} = -y\mathrm{sat}(y) \leq 0 \qquad (5.22)$$

We leave it to the reader to verify that the analysis based on LaSalle's Theorem produces the identical invariant set to which the trajectories of the closed loop system converge.

> **Experiment 16.** Implement the above energy/passivity controller on the Reaction Wheel Pendulum. What is the appropriate value of E_{ref} to use for swingup? Experiment with different values for the gains k_e, k_v, and k_u. *Note: It is better to do this in simulation before you try it out on the real system.* Plot the energy, pendulum angle, pendulum velocity, and disk velocity. How close do the actual values match the predicted values? Note also, that you may have to give the pendulum a slight push to get it started. Explain the reason for this. Explain why the pendulum comes close to the inverted position but does not balance there.

5.3.1 Some Experimental Results

The following plots (Figures 5.7–5.10) show the results of one such experiment on swingup control. We set the reference energy, E_{ref}, equal to the rest energy of the system in the inverted configuration. From the expression for the energy, we see that this corresponds to $E_{ref} = 2mg\ell$. With this value for E_{ref} the response of the system is shown below. The gains, k_e, k_v, and k_u were chosen as

$$k_e = 4000 \qquad k_v = 4 \qquad k_u = 0.4$$

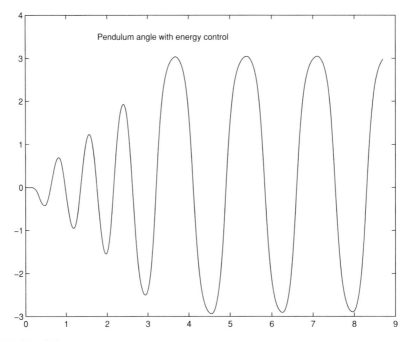

FIGURE 5.7: Pendulum response.

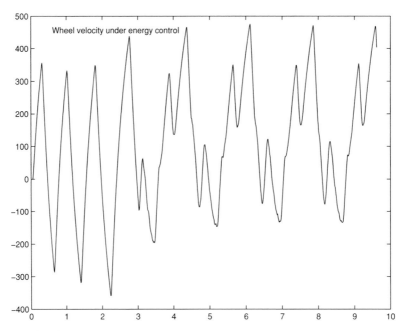

FIGURE 5.8: Disk velocity.

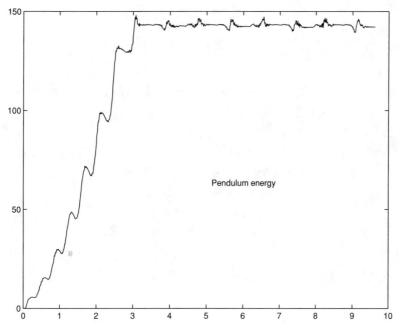

FIGURE 5.9: Pendulum energy.

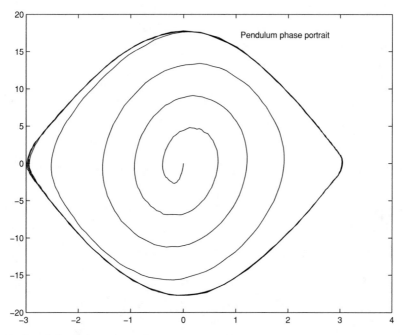

FIGURE 5.10: Pendulum phase portrait.

Summary

In this chapter, we have investigated the application of nonlinear control theory to the problem of swinging the pendulum up from the downward position to the inverted position. The concept of Passivity was shown to be a powerful design tool for the design of controllers for nonlinear systems.

Using Passivity together with LaSalle's Invariance Principle allowed us to present a complete analysis of the behavior of the closed loop system. We saw that, for almost all initial conditions, the pendulum trajectory would approach the inverted position and the wheel velocity would approach zero.

This control method can swing up the pendulum but cannot be used to balance it about the inverted configuration. For that we must combine the swingup control with a separate balance (or stabilization) control. The problem of how to switch between different controllers is considered next.

CHAPTER 6

Switching Control

6.1 SWINGUP COMBINED WITH STABILIZATION

So far we have designed linear control strategies to control the pendulum hanging down and standing up. These strategies also keep the wheel speed or wheel angle close to desired values. We have also designed nonlinear strategies for swinging up the pendulum. In this section, we will combine the different strategies. This leads to *Switching Controls*. The goal is to find a control strategy that keeps the pendulum upright. Since a large disturbance can always make the pendulum to swing down the strategy should have the property that it can recover even from large disturbances.

The energy control strategy has the form

$$u = k_u(k_e \dot{\theta}(E_{ref} - E) - k_v \dot{\theta}_r) \qquad (6.1)$$

where E is the scaled energy defined as

$$E = 1 - \cos \theta + \frac{1}{2} \left(\frac{\dot{\theta}}{\omega_p} \right)^2 \qquad (6.2)$$

the reference energy E_{ref} is zero when the pendulum is hanging down and $2mg\ell$ when the pendulum is upright. This energy control strategy brings the energy of the pendulum to the reference value E_{ref}. At the same time it also brings the wheel speed to zero.

The linear control strategies all have the form

$$u = -k_{pp}\theta - k_{dp}\dot{\theta} - k_{pr}\theta_r - k_{dr}\dot{\theta}_r \qquad (6.3)$$

where θ_{ref} is the desired orientation of the pendulum, $\theta_{ref} = 0$ when the pendulum is down, and $\theta_{ref} = \pi$ when it is up.

The energy controller moves the pendulum near the inverted configuration for (almost all!) initial conditions but cannot balance it there. The linear control strategies can stabilize the pendulum at the inverted position with zero wheel velocity but are only local. Clearly, then, a good strategy is to use the energy control until the pendulum is "close" to the inverted position and the "switch" control to the local stabilizing control. It turns out that this is not as easy as it may appear at first glance and there are a number of interesting issues associated with such switching controllers.

6.2   AVOIDING SWITCHING TRANSIENTS

It is desirable to avoid transients when switching between the controllers. To do this we will choose the stabilizing controller so that the vector field of the closed loop system with the stabilizing controller matches the homoclinic orbit when the pendulum is upright.

The energy controller drives the pendulum to a homoclinic orbit given by $E = 2$, which implies that

$$\frac{1}{2}\left(\frac{\dot{\theta}}{\omega_p}\right)^2 + \frac{1}{2}(1 - \cos\theta) = 2$$

Solving for $\dot{\theta}$ gives

$$\dot{\theta} = \omega_p\sqrt{2(1 - \cos(\theta - \pi))} \approx \pm(\theta - \pi)$$

The pendulum thus approaches the upright position along the straight line

$$\frac{\dot{\theta}}{\omega_p} = -\theta - \pi$$

In addition, the energy control strategy drives the wheel velocity to zero.

We will now investigate the vector field of the closed loop system obtained with the stabilizing strategy and determine the conditions required for the vector field to line up with the homoclinic orbit close to the upright position. Since energy control has feedback from E, $\dot{\theta}$, and $\dot{\theta}_r$, it is natural to have a stabilizing strategy of the form

$$u = -k_{pp}\theta - k_{dp}\dot{\theta} - k_{dr}\dot{\theta}_r$$

The linearized equation for the closed loop system with this control law is

$$\frac{dx}{dt} = \begin{pmatrix} 0 & 1 & 0 \\ b_p k_{pp} + a & b_p k_{dp} & b_p k_{dr} \\ -b_r k_{pp} & -b_r k_{dp} & -b_r k_{dr} \end{pmatrix} x = Ax$$

The trajectories of this system will match the homoclinic orbit if the matrix A has an eigenvalue ω_p associated with the eigenvector

$$e = \begin{pmatrix} 1 \\ -\omega_p \\ 0 \end{pmatrix}$$

Notice that the swing up strategy ideally comes in with zero wheel velocity. Hence

$$-\dot{\omega}_p = \lambda$$

$$b_p k_{pp} + a - b_p k_{dp}\omega_p = -\lambda\omega_p$$

$$-b_r k_{pp} + b_r k_{dp}\omega_p = 0$$

This implies that $\lambda = -\omega_p$ and that $k_{pp} = -\omega_p k_{dp}$.

We have thus found that the condition required for patching the vector fields obtained with energy control and stabilization is that

$$k_{pp} = -\omega_p k_{dp} \tag{6.4}$$

This condition implies that the eigenvalue of the closed loop system which is lined up with the homoclinic orbit is $-\omega_p$. This means that the pendulum will approach the equilibrium exponentially with rate $-\omega_p$. Also, notice that the condition (6.4) implies that the control signal is zero along the eigenvector.

The homoclinic orbit is curved but the orbit of the closed loop system with the stabilizing strategy is a straight line, namely the eigen-subspace. Figure 6.1 shows the orbit for the homoclinic orbit and the lines where the control signal is zero and at the saturation limits. To ensure that the trajectories with energy control and stabilization line up we thus have to impose

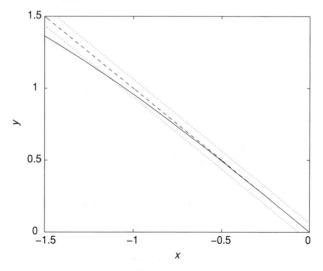

FIGURE 6.1: Trajectory for energy control (solid) and stabilizing control (dashed) with controller parameter chosen to patch the vector fields close to the origin. The dashed curves show the limits of the region where the stabilizing controller saturates.

the condition

$$|\theta - \pi| < \theta_s \qquad (6.5)$$

The figure shows that the trajectories are quite close for deviations from the upright position less than 0.5 rad, and we therefore choose $\theta_s = 0.5$. Notice that to get a fast swing up it is desirable to switch from the homoclinic orbit as soon as possible because the motion along the homoclinic orbit becomes very slow as it approaches the upright position.

6.3   FINDING PARAMETERS OF THE STABILIZING STRATEGY

Patching the vector fields of the energy controlled and the stabilized systems gives the condition (6.4). We will now discuss how the controller parameters should be chosen to satisfy this condition and to give other desirable properties. The characteristic polynomial of the system matrix A of the closed loop system is

$$A(s) = s^3 + (-b_p k_{dp} + b_r k_{dr})s^2 + (a - b_p k_{pp})s + a b_r k_{dr}$$

With $k_{pp} = -\omega_p k_{dp}$ the polynomial $A(s)$ has a root $s = -\omega_p$, and we have

$$A(s) = (s - \omega_p)(s^2 + (-b_p k_{dp} + b_r k_{dr} - \omega_p)s - b_r k_{dr}\omega_p)$$

Requiring that the second factor equals the standard second-order polynomial

$$s^2 + 2\zeta\omega_0 s + \omega_0^2$$

we find

$$-b_p k_{dp} + b_r k_{dr} - \omega_p = 2\zeta\omega_0$$
$$b_r k_{dr}\omega_p = -\omega_0^2$$

This gives

$$k_{pp} = -\frac{\omega_0^2 + 2\zeta\omega_0\omega_p + \omega_p^2}{b_p}$$

$$k_{dp} = -\frac{\omega_0^2 + 2\zeta\omega_0\omega_p + \omega_p^2}{b_p\omega_p}$$

$$k_{dr} = -\frac{\omega_0^2}{b_r\omega_p}$$

The controller parameters can be computed by the following Matlab program

```
function [kpp,kdp,kdr]=pddcontrolwpatch(w0,zeta,a,bp,br)
%Computation of feedback gains for stabilization of pendulum
%and control rotor speed for system with patching of energy
%and stabilization control
wp=sqrt(a);
kpp=-(w0^2+2*zeta*w0*wp+wp^2)/bp;
kdp=-(w0^2+2*zeta*w0*wp+wp^2)/bp/wp;
kdr=-w0^2/br/wp;
```

Choosing $\omega_0 = \omega_p$ and $\zeta = 0.707$ gives $k_{pp} = -248$, $k_{dp} = -28.0$, and $k_{dr} = -0.0447$. The choice $\omega_0 = 1.5 * \omega_p$ gives $k_{pp} = -389$, $k_{dp} = -44.0$, and $k_{dr} = -0.101$ and $\omega_0 = 2 * \omega_p$ gives $k_{pp} = -568$, $k_{dp} = -64.1$, and $k_{dr} = -0.179$.

6.4 SWITCHING CONDITIONS

So far we have given conditions for switching from energy control to stabilization. We must also give conditions for switching from stabilization to energy control. To do this we will first discuss when the saturated linear strategy can stabilize the pendulum. Assuming that the linear model is sufficiently accurate and assuming that the wheel velocity is sufficiently small it can be shown that a linear strategy that is patched to the homoclinic orbit will stabilize the system if

$$-\frac{b_p u_{max}}{\omega_p} - \omega_p \theta < \dot{\theta} < \frac{b_p u_{max}}{\omega_p} - \omega_p \theta \qquad (6.6)$$

where u_{max} is the largest control signal. Notice that the region of linear operation

$$-\frac{u_{max}}{|k_{dp}|} - \omega_p \theta < \dot{\theta} < \frac{u_{max}}{|k_{dp}|} - \omega_p \theta \qquad (6.7)$$

is much smaller. When the state goes outside the region (6.6) the pendulum will quickly fall down.

We will, therefore, simply switch to energy control when the pendulum angle deviates from the upright position by more than $\theta_0 = 0.5$ rad.

To illustrate that the conditions are reasonable, Figure 6.2 shows the projection of the trajectories on the θ-$\dot{\theta}$ plane for the tapping experiments in Figures 4.1 and 4.3. In the figure we have also shown the regions where the controller operates linearly and where it is able to maintain stability. Notice that the linear region is smaller for the controller with larger gains but that the regions where stability can be maintained is the same for both controllers. The figure indicates that the simple switching criteria is quite reasonable.

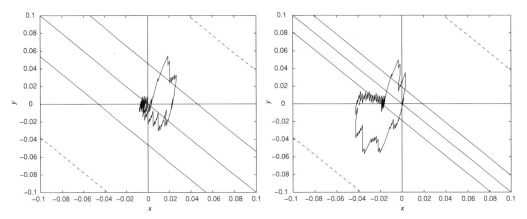

FIGURE 6.2: Results of tapping experiment with stabilization of a pendulum using PD control of pendulum and control of wheel speed. The controller was designed to give $\omega_0 = \omega_p$, $\zeta = 0.707$, and $\alpha = 0.2$. The controller parameters are $k_{pp} = -282$, $k_{dp} = -24.5$, and $k_{dr} = -0.0302$.

6.5 EXPERIMENTS WITH SWINGUP AND CATCHING
Figure 6.3 shows results of one experiment with swingup and catching of the pendulum. The catching strategy is designed to match the homoclinic orbit. Figure 6.4 shows the energy, the angular velocity of the pendulum, and the control signal. Notice that the energy does not

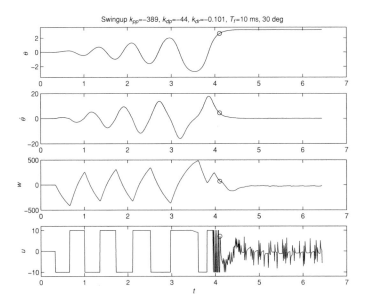

FIGURE 6.3: Experiment with swingup based on energy control and stabilization with a linear strategy which is matched to the homoclinic orbit, the controller gains for the linear strategy are $k_{pp} = -389$, $k_{dp} = -44$, and $k_{dr} = -0.101$. The angular velocity of the pendulum is filtered with a first-order filter having a time constant of 10 ms.

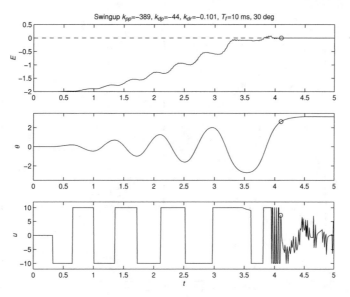

FIGURE 6.4: Plot of the energy of the pendulum in Figure 6.3. The curves given from above are scaled energy E, $\dot{\theta}$, and u.

decrease monotonically because the energy control strategy also attempts to keep the wheel velocity small. Figure 6.5 shows the projection of the trajectory on the $\dot{\theta}$-θ plane. The figure shows that the system behaves as intended and that the trajectories seem to match quite well during the switching. A more detailed picture which shows the catching phase is shown in

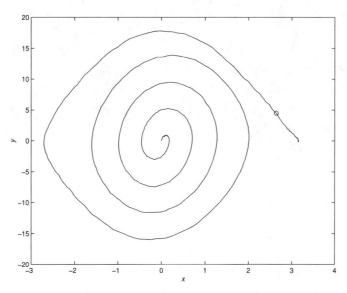

FIGURE 6.5: Projection of trajectories on the $\dot{\theta}$-θ plane for the data in Figure 6.3.

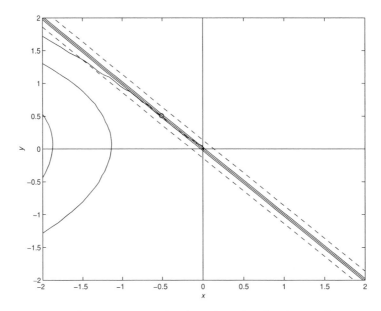

FIGURE 6.6: Projection of trajectories on the $\dot{\theta}$-θ plane for the data in Figure 6.3 zoomed at the catching zone.

Figure 6.6. In this figure, we have also shown the switching condition. The figure indicates that it is possible to catch the pendulum earlier than $\theta_s = 0.5$.

Summary

In this chapter, we have considered the problem of Switching Control in order to switch between swingup and stabilization for the problem of swinging the pendulum up to the inverted position and catching it there. We showed, by matching the trajectories of the swingup and balance controllers, how to eliminate the transient resulting from the switching.

CHAPTER 7

Additional Topics

7.1 AN OBSERVER FOR THE PENDULUM VELOCITY

We have found that the limited resolution of the encoder for the pendulum angle causes jumps in the control signal. We will now investigate if we can reduce the jumps by introducing an observer for the pendulum velocity. Since the only coupling between the pendulum and the wheel is via the control signal we can design an observer for the pendulum separately. The nonlinear equations of motion of the pendulum can be written in standard state space form as

$$\frac{dx_1}{dt} = x_2$$

$$\frac{dx_2}{dt} = a \sin x_1 - b_p(u - F)$$

$$y_1 = x_1$$

where $x_1 = \theta$ and $x_2 = \dot{\theta}$. The system is nonlinear but this is not serious since the signal x_1 is measured. An observer for the system is

$$\frac{d\hat{x}_1}{dt} = \hat{x}_2 + k_1(y - \hat{x}_1)$$

$$\frac{d\hat{x}_2}{dt} = a \sin y - b_p(u - \hat{F}) + k_2(y - \hat{x}_1)$$

where $\hat{F}$ is an estimate of the friction torque. This estimate can be obtained from the static friction model given by Eq. (3.17) or from the observer for the wheel angle, see Section 3.4. The observer error $e = x - \hat{x}$ is given by the equation

$$\frac{de_1}{dt} = -k_1 e_1 + e_2$$

$$\frac{de_2}{dt} = -k_2 e_1$$

This equation is linear and its characteristic polynomial is

$$s^2 + k_1 s + k_2$$

Identifying this with the standard second-order polynomial

$$s^2 + 2\zeta\omega_o + \omega_o^2$$

we find

$$k_1 = 2\zeta\omega_o$$
$$k_2 = \omega_o^2$$

The design of the observer is thus straightforward.

7.1.1   Sampling the Observer

A controller defined by a differential equation cannot be implemented directly using a computer. A difference equation is needed for computer implementation. To obtain this we approximate the derivative by a forward difference and we find that the observer can be described by

$$\hat{x}_1(t + h) = \hat{x}_1(t) + h\big(\hat{x}_2(t) + k_1(y(t) - \hat{x}_1(t))\big)$$
$$\hat{x}_2(t + h) = x_2(t) + h\big(a\sin y(t) - b_p(u(t) - \hat{F}(t)) + k_2(y(t) - \hat{x}_1(t))\big)$$

(7.1)

It now remains to find a reasonable value of the parameter ω_o that determines the bandwidth of the observer. The trade-offs are that a small value gives good filtering but slower response and vice versa. It is essential to have a reasonably fast observer during the swing up when the pendulum velocity can be quite large. In experiments with filtering of the velocity we found that a time constant $T = 10$ ms was reasonable. With this time constant we found that the jump in the filtered velocity caused by one encoder increment is

$$\Delta_{\dot{\theta}} = \frac{\Delta_\theta}{T} = 100\Delta_\theta$$

With the observer the equivalent jump in the estimate is

$$\Delta_{\dot{\theta}} = hk_2\Delta_\theta = h\omega_o^2\Delta_\theta$$

To have the same jump as with the velocity filter we must require that $k_2 < 1/hT = 100000$. With the values $h = 0.001$ and $T = 0.01$ we get $\omega_0 < 316$, smaller values of ω_o gives less ripple that with the filtered velocity difference.

> **Experiment 17 (Stabilization Using an Observer).**   Program the observer and make an experiment with feedback from the observed pendulum velocity. Investigate the properties of the system for different specifications on the observer, for example, by changing ω_o. Test the system by tapping it. Compare with the results where the pendulum velocity is determined using the filtered velocity difference.

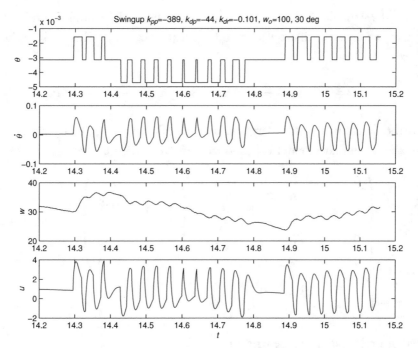

FIGURE 7.1: Stabilization of the pendulum with a linear controller having parameters $k_{pp} = -389$, $k_{dp} = -44$, and $k_{dr} = -0.101$. The angular velocity of the pendulum is determined using an observer with $\omega_o = 100$ and $\zeta = 0.7$.

Figure 7.1 shows results of an experiment with stabilization with feedback and an observer. The parameters of the observer were chosen as $\zeta = 0.707$ and $\omega_o = 100$. Using the formulas given above we find that the jumps in the velocity caused by the limited encoder resolution have the magnitude $h\omega_o^2 \Delta_\theta = 0.016$. Figure 7.1 also shows that the jumps are not noticeable in the velocity estimate and in the control signal. Compare with Figure 4.1 where the velocity was determined by filtering the velocity difference. As a result the control signal is also much smoother. There is, however, a limit cycle which is caused by the quantization of the rotor angle.

The observer with $\omega_o = 100$ gives a smoother control signal. The price of this is that the controller does not react as aggressively to disturbances. This can be investigated by tapping the pendulum.

7.1.2 Estimation of Friction

In Section 3.4, it was shown that the friction force can be determined using an observer. In Section 3.7, it was shown that the friction could be determined from the velocity difference of the wheel and the pendulum. Since friction influences both the pendulum and the wheel it is natural to make an observer that estimates all states jointly. To do this we introduce the state variables

$$x_1 = \theta, \quad x_2 = \frac{\dot{\theta}}{\omega_p}, \quad x_3 = \theta_r, \quad x_4 = \frac{\dot{\theta}_r}{\omega_p}, \quad x_5 = F$$

Notice that the derivatives of the states have been scaled with ω_p, this gives dimension free variables and a system that is better conditioned numerically. The state equations then become

$$\frac{dx}{dt} = \begin{pmatrix} 0 & \omega_p & 0 & 0 & 0 \\ \omega_p & 0 & 0 & 0 & b_p/\omega_p \\ 0 & 0 & 0 & \omega_p & 0 \\ 0 & 0 & 0 & 0 & -b_r/\omega_p \\ 0 & 0 & 0 & 0 & 0 \end{pmatrix} x + \begin{pmatrix} 0 \\ -b_p/\omega_p \\ 0 \\ b_r/\omega_p \\ 0 \end{pmatrix} u = Ax + Bu$$

$$\tag{7.2}$$

$$y = \begin{pmatrix} 1 & 0 & 0 & 0 & 0 \\ 0 & 0 & 1 & 0 & 0 \end{pmatrix} x = Cx$$

The observer for this system can be written as

$$\frac{d\hat{x}}{dt} = A\hat{x} + Bu + K(y - C\hat{x}) \tag{7.3}$$

where K is a 5×2 matrix. This case is borderline for analytical calculation and we will, therefore, use Matlab to compute the controller gains. To use pole placement we have to determine suitable poles for the observer. It is reasonable to have similar dynamics for estimation of the angular velocities and we therefore choose the characteristic polynomial of the observer as

$$(s^2 + 2\zeta\omega_o s + \omega_o^2)(s^2 + 2\zeta\omega_o s + \omega_o^2)(s + \alpha\omega_o)$$

The observer gains are then given by the following Matlab program

```
%Full state observer
%Computes the observer gains for an observer that estimates
%all states including friction
%systpar %Get system parameters
%A=[0 wp 0 0 0;wp 0 0 0 0;0 0 0 wp 0;0 0 0 0 -br/wp;0 0 0 0 0]
A=[0 wp 0 0 0;wp 0 0 0 bp/wp;0 0 0 wp 0;0 0 0 0 -br/wp;0 0 0 0 0];
B=[0;-bp/wp;0;br/wp;0];
C=[1 0 0 0 0;0 0 1 0 0];
P=roots([1 2*zeta*wo wo^2]);
```

```
P=[P;P;-alpha*wo];
K=place(A',C',P);
Kapprox=[2*zeta*wo wp+wo^2/wp 0 0 0;
0 0 (alpha+2*zeta)*wo (1+2*alpha*zeta)*wo^2/wp -alpha*wo^3/br];
```

Choosing $\omega_o = 100$, $\zeta = 0.707$, and $\alpha = 1$ we get

```
K'  = 1.0e+003 *
      0.1414     1.1381    -0.0005    -0.0087     0.0276
     -0.0005    -0.0087     0.2414     2.7259    -5.0505
Kapprox'  = 1.0e+003 *
      0.1414     1.1381          0          0          0
           0          0     0.2414     2.7259    -5.0506
```

Notice that several gains we have are very small

```
K'(1,3:5) = -0.5457     -8.7137     27.5624
K'(2,1:2) = -0.5457     -8.7137
```

The specifications $\omega_o = 100$, $\zeta = 0.707$, and $\alpha = 0.2$ imply that the estimation of friction is a little slower. With these data we get

```
K'  = 1.0e+003 *
      0.1414     1.1381    -0.0001    -0.0017     0.0055
     -0.0001    -0.0017     0.1614     1.4486    -1.0101
Kapprox'  = 1.0e+003 *
      0.1414     1.1381          0          0          0
           0          0     0.1614     1.4486    -1.0101
```

The specifications $\omega_o = 300$, $\zeta = 0.707$, and $\alpha = 1$ give

```
K'  = 1.0e+005 *
      0.0042     0.1017    -0.0000    -0.0008     0.0074
     -0.0000    -0.0008     0.0072     0.2453    -1.3636
Kapprox'  = 1.0e+005 *
      0.0042     0.1017          0          0          0
           0          0     0.0072     0.2453    -1.3637
```

The specifications $\omega_o = 300$, $\zeta = 0.707$, and $\alpha = 0.2$ give

```
K' = 1.0e+004 *
    0.0424    1.0172   -0.0000   -0.0016    0.0149
   -0.0000   -0.0016    0.0484    1.3037   -2.7272
Kapprox' = 1.0e+004 *
    0.0424    1.0172        0         0         0
        0         0    0.0484    1.3037   -2.7273
```

In the expressions above, we have also given the approximate filter gains obtained when the coupling between the systems have been neglected. The figures indicate that it is not necessary to consider the interaction between the systems when designing the observer. We can thus conclude that it is sufficient to use the simple observer given by Eq. (7.1) where the estimate of the friction is taken from the observer for the wheel. The performance can be improved by combining the friction observer with the friction model (3.17), which is particularly important in balancing where the friction changes sign frequently.

7.1.3   Swingup with an Observer

A nice property of the nonlinear observer (7.1) is that it works for the whole range of pendulum angles. It can, therefore, be used both in stabilization and swingup. We illustrate this with an experiment.

Experiment 18 (Swingup with an Observer).   Program the observer and make a swingup experiment when the velocities of the pendulum and the wheel are taken from the observer.

Figure 7.2 shows the results of an experiment with swingup of the pendulum where the observer is used to estimate the angular velocity of the pendulum. A comparison with Figure 6.3 shows that the observer gives a smoother control signal. In Figure 7.3, we show the projection of the trajectory on the $\dot{\theta}$-θ plane. The figure shows that the system behaves as intended and that the trajectories seem to match quite well during the switching. A comparison with the corresponding figure for the experiment when the velocity is determined by filtering the angular difference shows that the observer gives smoother trajectories. In Figure 7.4, we show the estimate of the pendulum velocity from the observer and from the filter. The estimates are quite similar but the figure shows that the estimate from the observer leads the filtered angle difference as can be expected. The difference is largest when the acceleration is the largest.

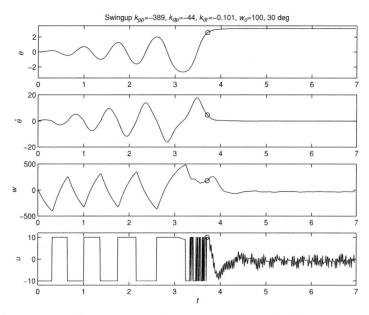

FIGURE 7.2: Experiment with swingup based on energy control and stabilization with a linear strategy which is matched to the homoclinic orbit, the controller gains for the linear strategy are $k_{pp} = -389$, $k_{dp} = -44$, and $k_{dr} = -0.101$. The angular velocity of the pendulum is determined using an observer with $\omega_o = 100$ and $\zeta = 0.707$.

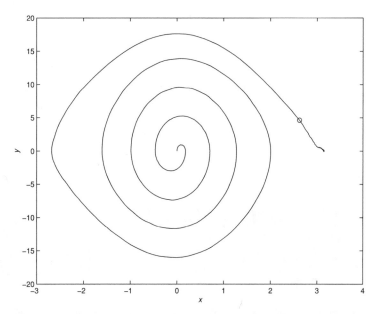

FIGURE 7.3: Projection of trajectories on the $\dot{\theta}$-θ plane for the data in Figure 7.2.

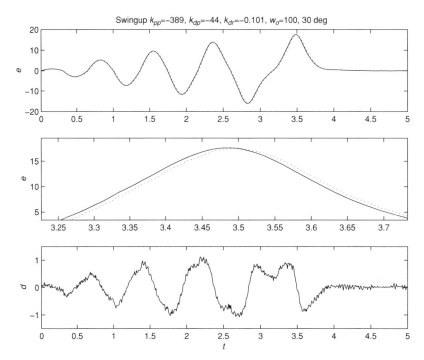

FIGURE 7.4: The upper curve shows the estimates of the velocity from the observer (solid) and the filtered difference (dashed). The middle curve is an enlargement of the upper curve and the lowest curve shows the difference $\hat{x}_2 - x_{2f}$.

7.2   MORE ABOUT FRICTION AND FRICTION COMPENSATION

When stabilizing the inverted pendulum we have so far neglected the friction force on the motor axis. We will now investigate the effects of the friction torque on the motor axis. The effects are most noticeable if the motor reverses direction frequently. We will start with an experiment using the control law

$$u = -k_{pp}\theta - k_{dp}\dot{\theta} - k_{pr}\theta_r - k_{dr}\dot{\theta}_r$$

> **Experiment 19 (Limit Cycles Generated by Friction).**   Stabilize the pendulum
> with the control law given by Eq. (4.13) with feedback from angles and velocities
> of pendulum and wheel. Observe the motion of the wheel. Explain qualitatively
> what happens. Record the motion of the wheel and characterize it.

Results of an experiment are shown if Figure 7.5. The figure shows that there is a limit cycle oscillation caused by the friction. The oscillation in the pendulum angle has small

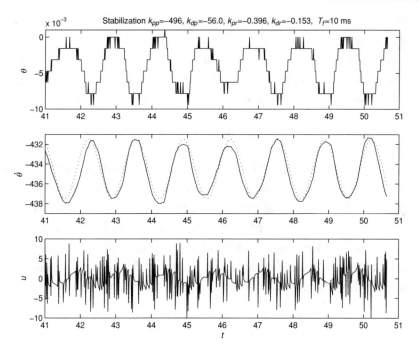

FIGURE 7.5: Stabilization of the pendulum with a linear controller having parameters $k_{pp} = -496$, $k_{dp} = -56$, $k_{pr} = -0.396$, and $k_{dr} = -0.153$. The angular velocity of the pendulum is determined by filtering the angle difference with a filter having time constant $T = 0.01$. The dotted curve shows a sinusoid with amplitude 3 and period 1.33 s.

amplitude corresponding to a few increments of the encoder. The oscillation in the pendulum angle is more sinusoidal. To analyze the oscillations we will use the describing function.

Mini tutorial on Describing Functions: *Describing functions* or the method of *harmonic balance* is an approximate method for determining limit cycles. It can be applied to a feedback loop composed of a linear and a nonlinear block as in Figure 7.6. The linear block is characterized

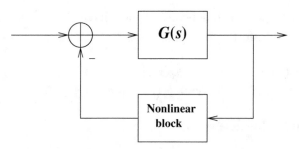

FIGURE 7.6: Closed loop system with one linear and one nonlinear block.

by its transfer function $G(s)$ and the nonlinear block is characterized by its describing function $N(a)$. The describing function characterizes the propagation of a sinusoidal signal through the system. The value $N(a)$ is a complex number that relates the first harmonic of the output to the input when the input is a sinusoid with amplitude a. Neglecting higher harmonics and tracing signals around the loop we find that an oscillation can be maintained if

$$G(i\omega)N(a) = -1 \tag{7.4}$$

This is an equation in two variables which can be solved by plotting the Nyquist curve of the transfer function $G(s)$ and the locus of $1/N(a)$. The solution is where the curves intersect. The values give approximately the frequency and the amplitude of a possible limit cycle. The criterion (7.4) is similar to Nyquist's stability criterion if the critical point is replaced by $-1/N(a)$.

It is also possible to investigate the stability of a limit cycle using describing functions. We define the direction of the describing function as the direction of increasing amplitude a. A point s_0 in the complex plane enclosed by the Nyquist curve of $G(i\omega)$ is called stable if

$$\frac{1}{2\pi}\Delta_\Gamma(s_0 + G(s)) + P = 0$$

where Δ_Γ is the argument variation on the Nyquist contour Γ (a half circle with the imaginary axis as diameter that encloses the right half plane) and P is the number of poles of $G(s)$ in the region enclosed by Γ. A potential limit cycle given by the intersection of the Nyquist curve with the describing function is stable if the direction of describing function at the intersection is toward the stable region.

End of Mini tutorial: To understand what happens in the experiment we will analyze the equations of the closed loop system. These equations can be written as

$$\ddot{\theta} = a\theta - b_p(u - F)$$
$$\ddot{\theta}_r = b_p(u - F)$$
$$u = -k_{pp}\theta - k_{dp}\dot{\theta} - k_{pr}\theta_r - k_{dr}\dot{\theta}_r$$

where F is the friction force, which depends on the relative rate of pendulum and wheel, i.e., $\dot{\theta} - \dot{\theta}_r$. Taking Laplace transforms the equation can be written as

$$\begin{pmatrix} s^2 - a - b_r(k_{dr}s + k_{dp}) & -b_p(k_{dr}s + k_{pr}) \\ b_r(k_{dr}s + k_{dp}) & s^2 - b_r(k_{dr}s + k_{pr}) \end{pmatrix} \begin{pmatrix} \Theta(s) \\ \Theta_r(s) \end{pmatrix} = \begin{pmatrix} b_p \\ -b_r \end{pmatrix} F(s)$$

The matrix on the left has the determinant

$$A(s) = s^4 + (-b_p k_{dp} + b_r k_{dr})s^3 + (-a - b_p k_{pp} + b_r k_{pr})s^2 - ab_r k_{dr}s - ab_r k_{pr}$$

and the equation has the solution

$$\begin{pmatrix} \Theta(s) \\ \Theta_r(s) \end{pmatrix} = \frac{1}{A(s)} \begin{pmatrix} s^2 - b_r(k_{dr}s + k_{pr}) & b_p(k_{dr}s + k_{pr}) \\ -b_r(k_{dr}s + k_{dp}) & s^2 - a - b_r(k_{dr}s + k_{dp}) \end{pmatrix} \begin{pmatrix} b_p \\ -b_r \end{pmatrix} F(s)$$

$$= \frac{1}{A(s)} \begin{pmatrix} b_p s^2 \\ -b_r(s^2 - a) \end{pmatrix} F(s)$$

hence

$$\Theta(s) - \Theta_r(s) = \frac{(b_r + b_p)s^2 - ab_r}{A(s)}$$

The transfer function from friction force F to velocity difference $\dot{\theta} - \dot{\theta}_r$ is then

$$G(s) = \frac{(b_r + b_p)s^3 - b_r a s}{s^4 + (-b_p k_{dp} + b_r k_{dr})s^3 + (-a - b_p k_{pp} + b_r k_{pr})s^2 - ab_r k_{dr}s - ab_r k_{pr}}$$

Assuming that friction is modeled by Coulomb friction we can determine possible limit cycles by describing function analysis. In Figure 7.7, we show the Nyquist curve of the transfer function G. Since the describing function for Coulomb friction is the negative real axis, describing function

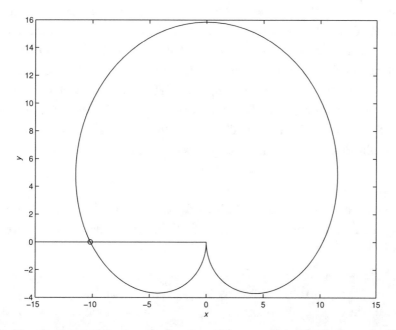

FIGURE 7.7: Nyquist curve of the transfer function G for a pendulum with the controller parameters $k_{pp} = -496$, $k_{dp} = -56.0$, $k_{pr} = -0.396$, and $k_{dr} = -0.153$.

analysis predicts a limit cycle at the intersection. The frequencies where $G(i\omega)$ intersects the real axis are given by

$$\omega^4 + (a + b_p k_{pp} + b_r k_{pr})\omega^2 - ab_r k_{pr} = 0$$

Inserting the numerical values we get $\omega^* = 4.12$, which corresponds to the period $T = 1.5$ s. Furthermore, we have $G(i\omega^*) = 10.2$. The describing function for Coulomb friction is

$$N(a) = \frac{4F_c}{a\pi}$$

where F_c is the friction level and a the amplitude of the oscillation in the angular frequency. Using the friction characteristics found in Section 3.7 we find that $F_c = 1$, see Figure 3.6.

The condition for oscillation is that

$$G(i\omega^*)N(a^*) = -1$$

We thus have the following estimate of the amplitude of the oscillation in the angular velocity

$$a^* = \frac{4F_c\, G(i\omega^*)}{\pi}$$

Inserting the numerical values we find $a^* = 13.0$. The corresponding amplitude of the angular oscillation is $a/\omega^* = 3.14$. The transfer function $G(s)$ is stable and the stability region is the interior of the curve in Figure 7.7. The direction of the describing function is toward the stability region and we can conclude that describing function theory predicts a stable limit cycle.

The describing function analysis indicates that the period of the oscillation is 1.5 s and that the oscillation of the wheel has an amplitude 3.14. The wheel will thus make substantial angular deviations, an oscillation with the amplitude 180°. A comparison with Figure 7.8 shows that the theory gives a reasonable prediction of the experimental results. A sinusoidal curve has been fitted to the wheel angle. This oscillation has the period 1.3 and the amplitude 3. The agreement with the describing function analysis is quite reasonable.

7.2.1   Friction Compensation

Since we have estimates of the velocity, it is straightforward to make a friction compensation by computing the friction using Eq. (3.17) and adding the signal to the control signal. We illustrate this by an example.

> **Experiment 20 (Friction Compensation).**   Repeat Experiment 19. Determine the size of the limit cycle by measuring the amplitude of the oscillations of the wheel. Switch in friction compensation and observe how much the amplitude of the oscillations are reduced.

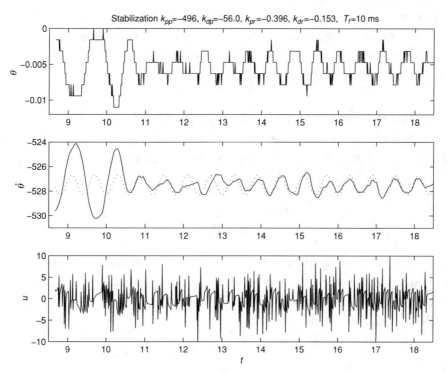

FIGURE 7.8: Stabilization of the pendulum with a linear controller having parameters $k_{pp} = -496$, $k_{dp} = -56$, $k_{pr} = -0.396$, and $k_{dr} = -0.153$. The angular velocity of the pendulum is determined by filtering the angle difference with a filter having time constant $T = 0.01$. Friction compensation is switched in at time 10.5. The dotted curve shows a sinusoid with amplitude 0.8.

Figure 7.8 shows the results obtained with friction compensation. The system is first run with the linear controller. The friction compensation is switched in at time 10.5. Notice that there is a substantial reduction in the amplitude of the limit cycle. Without friction compensation the amplitude of the wheel angle is about 3 rad, with friction compensation is reduced to 0.8 rad. The frequency of the oscillation is also increased. Compare also with Figure 7.5. The friction characteristics will change with many factors. To be effective, friction compensation should therefore be adaptive.

7.3 QUANTIZATION

In practical, all the experiments we have seen are effects of the encoder resolution. We will now investigate some consequences of this in more detail. In particular, we will study the possible limit cycles when the pendulum is stabilized in the upright position. For this purpose, we will approximate the resolution with a quantizer. The phenomena we are looking are clearly nonlinear. The experiments indicate that there are differences between the cases where the pendulum velocity is determined by filtering the angle increments and when an observer is used. Compare the stabilization experiments in Figures 4.4 and 7.1. There is clearly a qualitative difference between these two figures. Let us see if we can explain it.

7.3.1   Velocity Estimates by Filtered Angle Increments

We will first investigate the case when the velocity is obtained by filtering the velocity increments. Neglecting quantization and filtering, the closed loop system is described by

$$\ddot{\theta} = a\theta - b_p u$$
$$\ddot{\theta} = b_r u$$
$$u = -k_{pp}\theta - k_{dp}\dot{\theta} - k_{dr}\dot{\theta}_r$$

Based on the results of the experiments we will neglect the quantization in the wheel angle. We will first consider the case when the angular velocity of the pendulum is determined by filtering the increment in pendulum angle over a sampling interval.

To describe the filtering we introduce

$$V(s) = (k_{pp} + k_{dp}H(s))Y(s) = K(s)Y(s)$$

where y is the encoder signal from the pendulum angle and H is the transfer function of the filter used to generate the velocity. For a simple first-order filter we have

$$H(s) = \frac{s}{1 + sT}$$

Taking Laplace transform of the above equation we find

$$(s^2 - a)\Theta(s) = b_p K(s)Y(s) + b_p k_{dr}s\,\Theta_r(s)$$
$$s^2\Theta_r(s) = -b_r K(s)Y(s) - b_r k_{dr}s\,\Theta_r(s)$$

Eliminating $\Theta_r(s)$ we find

$$\Theta_r(s) = \frac{b_p s\,K(s)}{(s^2 - a)(s + b_r k_{dr})}Y(s)$$

This nonlinear equation can be interpreted as the feedback connection of the transfer function

$$G(s) = -\frac{b_p s\,K(s)}{(s^2 - a)(s + b_r k_{dr})} = -\frac{b_p s(k_{pp} + k_{dp}H(s))}{(s^2 - a)(s + b_r k_{dr})}$$

and a quantizer. We will investigate if the equation has a limit cycle using the approximate describing function method.

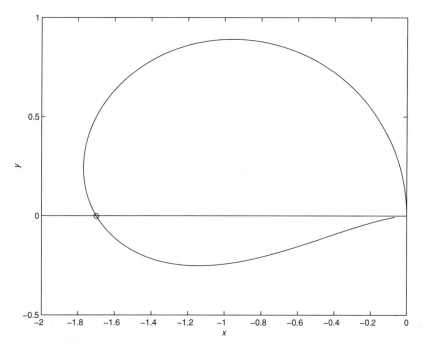

FIGURE 7.9: Nyquist curve of the transfer function G for a pendulum with the controller parameters $k_{pp} = -282$, $k_{dp} = -24.5$, $k_{pr} = -0.0302$ where the velocity is obtained by filtering the angle difference by a first-order filter with time constant 0.02 s.

With a first-order filter we have

$$G(s) = -\frac{b_p s \big((k_{pp} T + k_{dp})s + k_{pp}\big)}{(s^2 - a)(s + b_r k_{dr})(1 + s\, T)}$$

Figure 7.9 shows the Nyquist curve of the transfer function $G(s)$. The Nyquist curve is generated by the following Matlab program

```
%Analysis of effects of quantization for stabilizing controller
%with velocity estimate formed by filtering angle increments
systpar;
T=0.02;
kpp=-389;
kdp=-44;
kdr=-0.101;
num=-bp*[kpp*T+kdp kpp 0]
den=conv([1 0 -a],[1 br*kdr]);
```

```
den=conv(den,[T 1])
w=0.01:0.1:200;
[re,im]=nyquist(num,den,w);
[i,j]=min(abs(im));
x=0:0.1:2;
reax=zeros(size(x));
%plot(re,im,re(j),im(j),'ro',-x,reax,re,-im,'b--')
%gives complete Nyquis contour
plot(re,im,re(j),im(j),'ro',-x,reax)
axis([-2 0 -0.5 1])
rec=re(j)
imc=im(j)
wosc=w(j)
```

The describing function of a quantizer normally given in the literature assumes that the input is exactly in the middle of the quantization interval. The describing function is then zero for amplitudes less than half a quantization interval. It increases rapidly toward a maximum of 1.27 and it then oscillates toward the value one for large values of a. If the input signal is centered at a quantization step, the describing function is equal to the describing function for a relay, i.e.,

$$N(a) = \frac{4}{a\pi}$$

The transfer function $G(s)$ has two poles in the right half plane. The argument variation is zero and the drop-shaped region is thus an unstable region. The describing function points toward the unstable region and describing function theory predicts that the potential limit cycle is unstable. Describing function theory thus predicts that there will not be a limit cycle due to the limited resolution of the encoder. The experiments as in Figure 7.1 indicate that there are irregular motions in the pendulum angle but not limit cycles. The irregular motion is caused by disturbances. Because of the sensor resolution there will be no feedback if the motion of the pendulum is less than one encoder resolution. Disturbances combined with the instability of the pendulum will cause the motion.

7.3.2   Velocity Estimates from an Observer

We will now investigate the case when the velocity of the pendulum is obtained form an observer. We will make the same assumptions as in the previous case, namely that the quantization of the wheel angle can be neglected. The closed loop system can be described by the equations

$$\ddot{\theta} = a\theta - b_p u$$

$$\ddot{\theta}_r = b_r u$$

$$\frac{d\hat{\theta}}{dt} = \dot{\hat{\theta}} + k_1(y - \hat{\theta})$$

$$\frac{d\dot{\hat{\theta}}}{dt} = a\theta + b_p u + k_2(y - \hat{\theta})$$

$$u = -k_{pp}y - k_{dp}\dot{\hat{\theta}} - k_{dr}\dot{\theta}_r$$

Introducing the state variables $x_1 = \theta$, $x_2 = \dot{\theta}$, $x_3 = \theta_r$, $x_4 = \dot{\theta}_r$, $x_5 = \hat{\theta}$, $x_6 = \dot{\hat{\theta}}$, the system can be described by the equations

$$\frac{dx}{dt} = \begin{pmatrix} 0 & 1 & 0 & 0 & 0 & 0 \\ a & 0 & 0 & b_p k_{dr} & 0 & b_p k_{dp} \\ 0 & 0 & 0 & 1 & 0 & 0 \\ 0 & 0 & 0 & b_r k_{dr} & 0 & b_p k_{dp} \\ 0 & 0 & 0 & 0 & -k_1 & 1 \\ 0 & 0 & 0 & b_p k_{dr} & -k_2 & b_p k_{dp} \end{pmatrix} x + \begin{pmatrix} 0 \\ b_p k_{pp} \\ 0 \\ -b_r k_{pp} \\ 0 \\ a + k_2 + b_p k_{pp} \end{pmatrix} y \qquad (7.5)$$

$$\theta = \begin{pmatrix} 1 & 0 & 0 & 0 & 0 & 0 \end{pmatrix}$$

where the encoder output y is a quantization of the encoder angle. To investigate if this system can exhibit a limit cycle we observe that Eq. (7.5) can be regarded as a feedback connection of a quantizer and a linear system with the transfer function

$$G(s) = -C(sI - A)^{-1}B$$

Figure 7.10 shows the Nyquist curve of the transfer function $G(s)$. The Nyquist curve is generated by the following Matlab program

```
%Analysis of effects of quantization for system
%with observer for pendulum velocity
%x1=theta,x2=ptheta,x3=thetar,x4=pthetar,x5=hattheta,x6=phattheta
%systpar;
systpar
w0=1.5*wp;
zeta=0.707;
alpha=0.2;
k1=140;
```

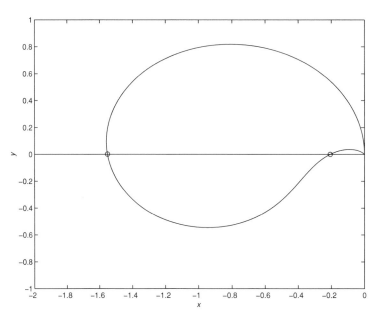

FIGURE 7.10: Nyquist curve of the transfer function G for a pendulum stabilized with a controller having the parameters $k_{pp} = -282$, $k_{dp} = -24.5$, $k_{pr} = -0.0302$, where the velocity is obtained by an observer with $\omega_0 = 100$ and $\zeta = 0.707$.

```
k2=10000;
[kpp,kdp,kdr]=pddcontrolwpatch(w0,zeta,a,bp,br)
%kpp=-389;kdp=-44,kdr=-0.101;
A=[0 1 0 0 0 0;
   a 0 0 bp*kdr 0 bp*kdp;
   0 0 0 1 0 0;
   0 0 0 -br*kdr 0 -br*kdp;
   0 0 0 0 -k1 1;
   0 0 0 bp*kdr -k2 bp*kdp]
B=[0;bp*kpp;0;-br*kpp;k1;a+k2+bp*kpp];
C=-[1 0 0 0 0 0];
D=0;
g=ss(A,B,C,D);
w=0.01:0.05:500;
[re,im]=nyquist(g,w);
re=re(:);
im=im(:);
[vi,j]=min(abs(im(1:500)));
```

```
[vm,m]=min(abs(im));
x=0:0.1:2;
reax=zeros(size(x));
plot(re,im,-x,reax,re(j),im(j),'ro',re(m),im(m),'ro')
axis([-2 0 -1 1])
gor=re(j)
goi=im(j)
wo=w(j)
```

Figure 7.10 shows that there are two intersections between the describing function for the relay and the Nyquist curve corresponding to frequencies 12 and 120 rad/s. The matrix A has two eigenvalues in the right half plane. The stable region is thus the interior of the contour and the intersection at the higher frequency corresponds to a stable limit cycle. Notice, however, that if the input to the describing signal is centered right between two encoder marks the describing function has the maximum value and there is no intersection between the Nyquist curve and the describing function. Depending on the average value of the pendulum angle describing function theory thus predicts that there is a limit cycle when the angle is close to a mark of the encoder but that there is no limit cycle when the angle is between two encoder marks. This explains the behavior shown in Figure 7.1 where there are intervals with and without limit cycles.

Summary

In this chapter, we have provided further discussion on both friction and the design of observers. We showed that a nonlinear observer for the pendulum angle can be designed such that the equations for the observer error states are linear. This interesting property of the Inertia Wheel Pendulum results from the fact that the only nonlinearity in the system is a function of the pendulum angle and hence is measurable. The observer in Section 7.1 is thus a special case of the use of so-called *Output Injection* [7].

We then discussed further the effects of sampling, quantization, and friction on the observer system and we showed the use of the observer together with the nonlinear swingup controller. We next used *Describing Functions* to analyze limit cycles due to friction. We have shown that friction can lead to limit cycle oscillations. We showed that a friction compensation based on a simple model can give a significant reduction of the amplitude of the limit cycle. We also discuss quantization, which is a strongly nonlinear phenomena and which is difficult to analyze. The approximate describing function method gives insight into the behavior of systems with quantization.

Bibliography

[1]   Åström, K.J., and Furuta, K., "Swinging up a pendulum by energy control," *Automatica*, vol. 36, pp. 278–285, 2000. doi:10.1016/S0005-1098(99)00116-8

[2]   Åström, K.J., "Hybrid control of inverted pendulums," In Y. Yamamoto and S. Hara, Eds., *Learning, Control, and Hybrid Systems*, pp. 150–163, Springer, New York, 1999.

[3]   Åström, K.J., Furuta, K., Iwashiro, M., and Hoshino, T., "Energy based strategies for swinging up a double pendulum," *14th World Congress of IFAC*, vol. M, pp. 283–288, Beijing, P.R. China, July 1999.

[4]   Chung, C.C., and Hauser, J., "Nonlinear control of a swinging pendulum," *Automatica*, vol. 31, pp. 851–862, 1995. doi:10.1016/0005-1098(94)00148-C

[5]   Fantoni, I., Lozano, R., and Spong, M.W., "Energy based control of the pendubot," *IEEE Transactions on Automatic Control*, vol. AC-45, no. 4, pp. 725–729, April 2000. doi:10.1109/9.847110

[6]   Gäfvert, M., Svensson, J., and Åström, K.J., "Friction and friction compensation in the Furuta pendulum," *Proceedings of the 5th European Control Conference (ECC'99)*, Karlsruhe, Germany, 1999.

[7]   Isidori, A., *Nonlinear Control Systems*, 3rd ed., Springer-Verlag, London, 1995.

[8]   Khalil, H., *Nonlinear Systems*, 2nd ed., Prentice-Hall, Upper Saddle River, NJ, 1996.

[9]   Lozano, R., Fantoni, I., and Block, D., "Stabilization of the inverted pendulum around its homoclinic orbit," *Systems and Control Letters*, vol. 40, pp. 197–204, 2000. doi:10.1016/S0167-6911(00)00025-6

[10]   Olfati-Saber, R., "Globally stabilizing nonlinear feedback design for the inertia-wheel pendulum," preprint, 2000.

[11]   Ortega, R., and Spong, M.W., "Stabilization of underactuated mechanical systems via interconnection and damping assignment," *IFAC Workshop on Lagrangian and Hamiltonian Methods for Nonlinear Control*, Princeton, NJ, March 16–18, 2000.

[12]   Shiriaev, A., Pogromsky, A., Ludvigsen, H., and Egeland, O., "On global properties of passivity-based control of an inverted pendulum," *International Journal of Robust and Nonlinear Control*, vol. 10, pp. 283–300, 2000. doi:10.1002/(SICI)1099-1239(20000415)10:4<283::AID-RNC473>3.0.CO;2-I

[13]   Spong, M.W., Block, D.J., and Astrom, K.J., "The mechatronics control kit for education and research," *IEEE Conference on Control Applications*, Mexico City,

Sept. 2001.

[14] Spong, M.W., "Passivity based control of the compass gait biped," *IFAC World Congress*, Beijing, China, July 1999.

[15] Spong, M.W., and Tsao, T.-C., "Mechatronics education at the University of Illinois," *IFAC World Congress*, Beijing, China, July 1999.

[16] Spong, M.W., and Praly, L., "Control of underactuated mechanical systems using switching and saturation," In *Proceedings of the Block Island Workshop on Control Using Logic Based Switching*, Springer-Verlag, London, 1996.

[17] Spong, M.W., and Block, D.J., "The pendubot: A mechatronic system for control research and education," *34th IEEE Conference on Decision and Control*, pp. 555–556, New Orleans, Dec. 1995.

[18] Spong, M.W., Corke, P., and Lozano, R., "Nonlinear control of the inertia wheel pendulum," *Automatica*, to appear.

[19] Spong, M.W., "Some aspects of switching control in robot locomotion," In *at-Automatisierungs Technik*, vol. 4, pp. 157–164, Oldenbourg Verlag, April 2000. doi:10.1524/auto.2000.48.4.157

[20] Spong, M., "The swingup control problem for the acrobot," *IEEE Control Systems Magazine*, vol. 15, no. 1, pp. 49–55, Feb. 1995. doi:10.1109/37.341864

[21] Spong, M.W., and Vidyasagar, M., *Robot Dynamics and Control*, John Wiley & Sons, New York, 1989.

[22] Wiklund, M., Kristenson, A., and Astrom, K.J., "A new strategy for swinging up an inverted pendulum," *Proceedings of the IFAC Symposium*, Sydney, Australia, 1993.

Index

Author Biography

Daniel J. Block is currently Manager of the College of Engineering Control Systems Laboratory at the University of Illinois at Urbana-Champaign. He also is the lecturer for two control systems laboratory courses. Dan received his Bachelor of Science degree and Master of Science degree in General Engineering at the University of Illinois in 1991 and 1996, respectively.

Karl J. Åström was educated at the Royal Institute of Technology (KTH) in Stockholm, Sweden. He worked for 5 years for IBM Research Laboratories in Stockholm, Yorktown Heights and San Jose. In 1965 he became Professor, Chair of Automatic Control at Lund Institute of Technology/Lund University where he built a new department. He is now is Emeritus at Lund University. From 2000 he is part time visiting professor at University of California in Santa Barbara. Åström has broad interests in automatic control covering both theory and applications. He has coauthored 10 books and more than 150 papers, one paper coauthored by Björn Wittenmark was selected for publication in the IEEE book Control Theory: Twenty-five seminal papers. Åström is listed in ISIHighlyCited, and he has several patents. One on auto-tuning coinvented by Tore Hägglund has led to significant manufacturing. Åström has received many awards including the 1993 IEEE Medal of Honor and the 1987 Quazza medal from International Federation of Automatic Control.

Mark W. Spong is currently Donald Biggar Willett Distinguished Professor of Engineering, Professor of Electrical and Computer Engineering, and Research Professor in the Coordinated Science Laboratory at the University of Illinois at Urbana-Champaign. He is also Director of the Center for Autonomous Engineering Systems and Robotics (CAESAR) within the Information Trust Institute (ITI) at Illinois. Dr. Spong's main research interests are in robotics, mechatronics, and nonlinear control theory. He has published more than 200 technical articles in control and robotics and is co-author of four books. Dr. Spong is Past President of the IEEE Control Systems Society and a Fellow of the IEEE. His recent awards include the Senior U.S. Scientist Research Award from the Alexander von Humboldt Foundation, the Distinguished Member Award from the IEEE Control Systems Society, the John R. Ragazzini and O. Hugo Schuck Awards from the American Automatic Control Council, and the IEEE Third Millennium Medal.